ARDUINO LED STRIP PROJECTS

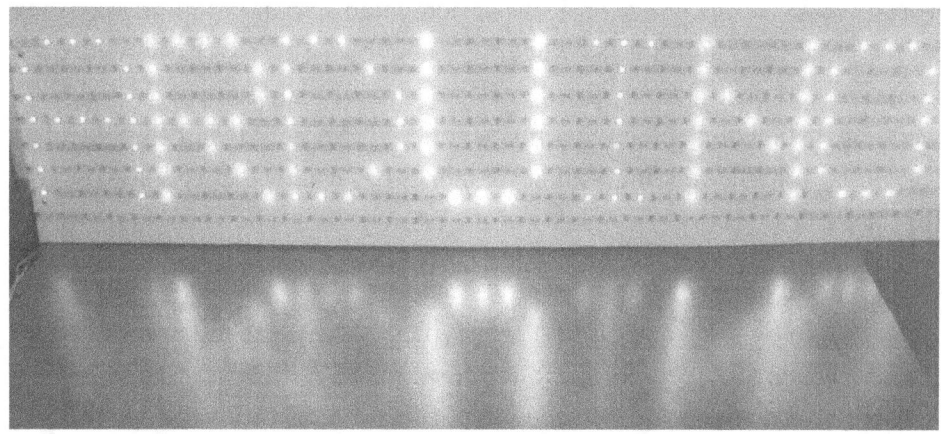

Robert J Davis II

Arduino LED Strip Projects

How to Build LED Signs with Addressable LED's

Copyright 2018 by Robert J Davis II

The safe construction and operation of the projects in this book are the sole responsibility of the reader or builder. Running these LED strips takes lots of power and can easily burn up voltage regulators. I speak from experience..

It was not easy to pick the title of this book. It started out with LED Strips but expanded to include other forms of addressable LED's. Should this book be just about the WS2812? Most people do not know what a WS2812 is! Should it be limited to LED Strips and not include other arrays? I could not come up with an all inclusive title for the book other than perhaps Addressable LED Projects. But Addressable was too long of a word to put in a book title.

These WS2812 LED Strip projects started when a friend of mine asked for a very large LED sign for his wedding reception. The sign started out at 40 inches long and quickly grew to 12 feet long by the time of the wedding. Then he decided that it was too big! So only two 4 foot panels were used for the wedding. Then someone that was at the wedding asked for the LED sign for their display at the county fair. They had a tent that was about 20 feet wide so they could easily us the entire sign.

Banks, High Schools and other places typically pay $10,000 or more for their LED signs. I recently saw a proposal for $37,000 for a LED sign! You can make your own LED signs for a fraction of the cost of buying one! This book will show you some of the possible arrangements of addressable LED strips and arrays and give you some examples of the software that will be need to make these arrangements work.

Introduction

1. Introducing LED Strips.. 4
    Analog or Digital Strips
    WS2812 Timing Considerations
    Power Requirements
    Parts Sources

2. Running One LED Strip or Array.. 14
    LED Strips
    8x8 Array
    Color and Worm Test Program
    15x16 Array
    8x32 Array
    Zigzag Text Program

3. Interfacing to Eight LED Strips... 30
    30 LED per Meter Sign
    60 LED per Meter Sign
    144 LED per Meter Sign
    Multicolor Text Sign Program
    Scrolling Text Sign Program

4. Adding a MSEQ7 for More Effects... 51
    Audio Spectrum Analyzer
    42 Frequency Spectrum Analyzer
    Oscilloscope Like Display

5. Adding Serial Communication... 67
    Bluetooth Adapter
    RF Wireless USB Adapter

6. Interfacing to 12 LED Strips.. 83
    Creating Fonts

7. Displaying Graphics.. 90

Bibliography..100

Chapter 1

Introducing LED Strips

Analog or Digital Strips?

The first LED strip that I ordered turned out to be an "Analog" LED strip. It consisted of red, green and blue LED's alternating down the strip. It is good for Christmas tree lights or for lighting up the arms of a drone so you can quickly tell where the front is, but it is generally used for accent lighting.

In fact there are many types of analog LED strips. One has red, green and blue LED's another has tri color LED's others have just white LED's. The advantage of having tri color LED's in that they can produce white light by lighting all three colors at the same time. However with an analog LED strip you cannot select individual LED's. If you turn blue on all the blue LED's light up. So it is not possible to program the individual LED's.

The WS2811 is addressable but the LED is usually separate. Usually these come with wires in between them so they look like Christmas tree lights. However they are addressable so you can do all kinds of effects with them as well. If you have something like a plastic star and want to drill holes in it and add addressable lights the WS2811 with the LED's would be what you would want to use. This next picture is of a typical string of 50 WS2811's with the three color LED's and it is sealed for outdoor use.

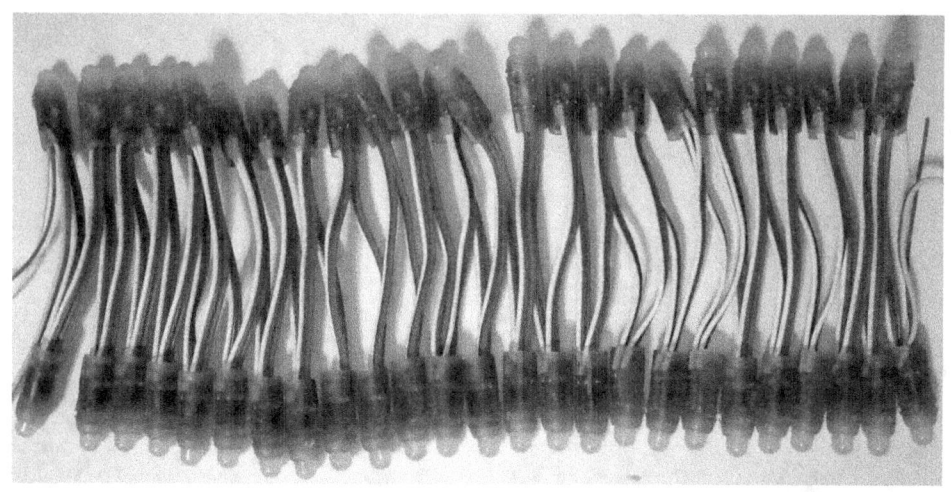

The WS2812 gives you the ability to address any one LED or lots of them at the same time. So you can make LED signs, effects, decorations, etc. There is almost no limit on what you can do with these addressable LED's.

Analog Strips: 3528 Red Green Blue LED
                5050 Tri Color LED
                3020 White LED

Digital Strips: WS2811+LED   Christmas Tree Lights
                WS2812   144 LED per Meter
                WS2812   60 LED per Meter
                WS2812   30 LED per Meter

The actual spacing of the LED's is about 1.75 inches for 30 LED per meter 7/8 inches for 60 LED per meter and a little over 1/4 inch for the 144 LED per meter strips. If you are making a sign those are the spacing's that you can use for marking the sign board before attaching the LED strips. The LED strips have a junction every 1/2 meter or about every 20 inches. So the ideal sign width is 20, 40, or 60 inches wide. You can make other widths by soldering together smaller segments. Like for a sign that is 48 inches wide using 30 LED per meter strips you would use two segments of 15 LED's and one of 6 LED's for a total of 36 LED's.

This next picture shows two 30 LED per meter strips on top, two 60 LED per meter strips in the middle and two 144 LED per meter strips at the bottom.

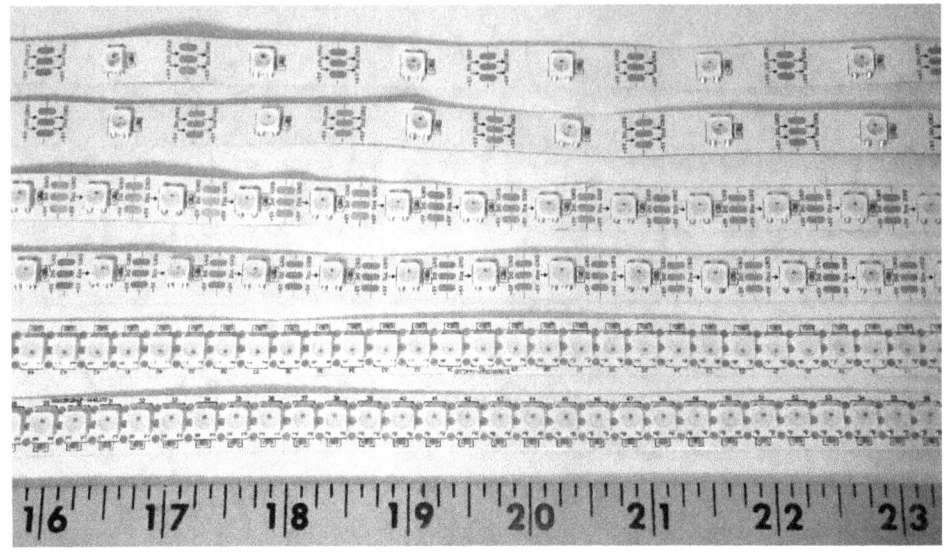

Note that there are the following options:

The WS2812 is also available in panels of 8x8, 16x16 or 8x32.
Available in one Meter (39.4 inches) to five meter lengths.
Actual LED strip segments are about 19.7 inches long.
Available in waterproof (Silicon coated) or normal.
Available in 5 volt and 12 volt power options.

If you are making LED signs or panels then you will also need some JST SM connectors. They are usually purchased with short wires already attached to make connecting them up to the LED strips easier.

LED strips have improved over the years. Earlier models had separate data and clock pins. Today they have only three connections. They are: Power or 5 volts, data (and clock combined) and ground. This makes it easy to hook them up, but it also places some timing limits on how many you can address in a period of time.

WS2812 Timing Considerations

The timing for the WS2812 is not as critical as some might think. All the timing is plus or minus 150ns. That is a range of .3us! This is the WS2812 timing diagram adapted from the specification sheet.

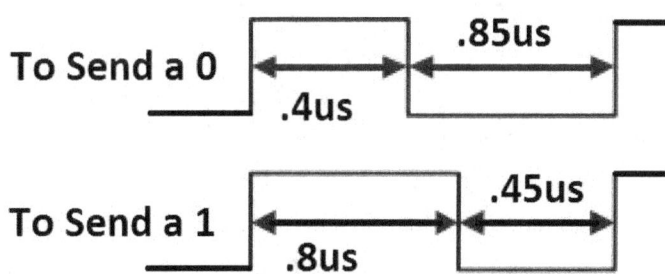

However that chart can be simplified into three blocks of .4us each. That is what I have done with the software. The first block is always high, the second block is high for a 1 and low for a 0 and the third block is low.

You cannot control the liming of the last low because it will be a while before the processor gets back to sending data. Just set all bits low and you have about .3us to figure out what to display next and get back to sending data. Accomplishing that is not easy with the Uno. You have to find the fastest possible way to do things.

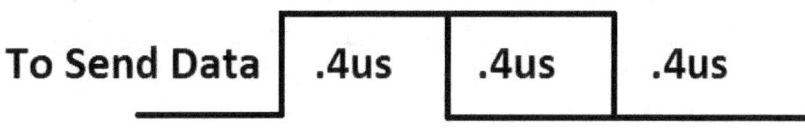

Each pixel needs 8 bits of red, 8 bits of green and 8 bits of blue for a total of 24 bits. The first pixel collects its needed 24 bits. Then all of the following bits are sent on to the second pixel. That second pixel collects the next 24 bits and then all of the following bits are sent on to pixel three. When there is a very long pause between bits, the pixels then display the data they have collected.

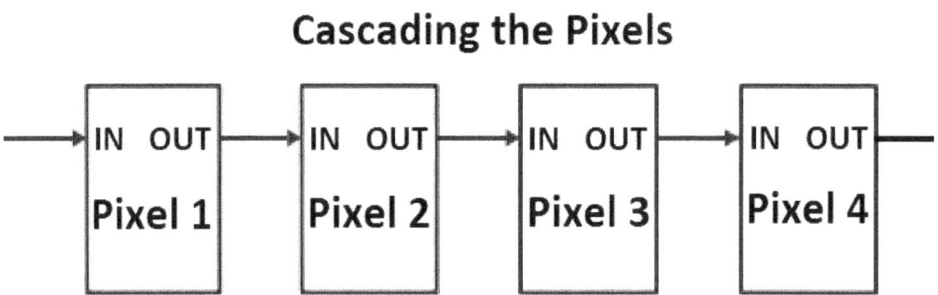

The biggest issue with the timing come when the processor has to do so much work that the display starts to timeout and will then display what was being sent to it. This happens the most when trying to display text characters. The processor has to look up each character, then find the byte within the character, then send that byte to be displayed in the sign. Sometimes those operations take too long.

I have tried to pass the color setting as three values: (red, green, blue). But passing those three variables, every time something is sent to the sign, may result in a "time out" situation where some of the WS2812's would think the processor was done sending things. This produces a garbage display.

Power Requirements

The power requirements of LED strips can be a serious problem. Let us say you are running one LED strip with 300 LED's. Each pixel requires .01 Amps for each of the three colors or .03 amps times 300 LED's. That is nine amps of power!

Displaying text does not take as much power because only one color is on at a time and only about 1/4 of the pixels are on at a time. Displaying one color at a time would be .01 amps times 300 LED's or three amps of power. I have done 16 rows of 90 LED's for 1440 LED's off a 12 amp power supply. But the power supply was making whining noises as it was a bit overloaded. One color turned on all 1440 pixels would be 14.4 amps.

So then a two amp five volt AC adapter can operate a small 8x8 LED array. For anything bigger that that I would recommend using a 12 amp or larger power supply.

This picture shows a typical five volt 12 amp power supply. The 120 volt wire coming in at the top need to be well insulated. I used some quick disconnect connectors left over from making a quadcopter power supply to make connecting and disconnecting the five volt output to the LED signs easier.

For mobile operation a 12 volt to 5 volt at 10 amps DC-DC converter will do the trick. I have used one of these DC-DC converters to run a LED sign at festivals for several hours off a car battery.

This next picture is of a 10 amp DC-DC converter. For full power operation it might need a heat sink to keep it cool. Large alligator clips clip onto the battery terminals. You could use a cigarette lighter adapter for a power source as well.

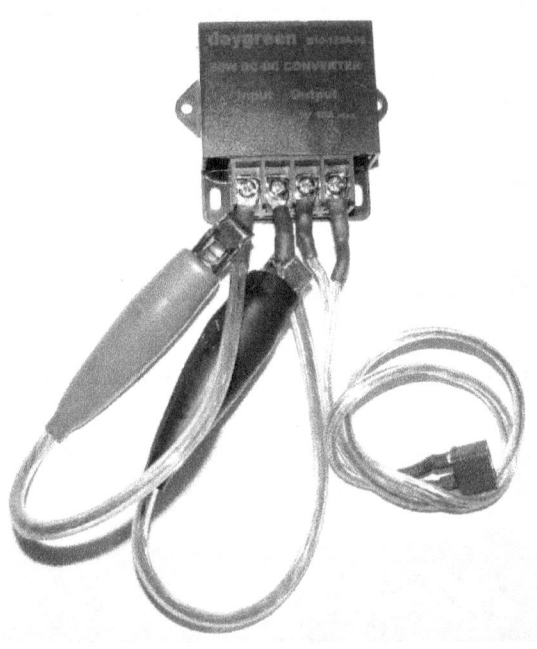

Another power option is to use a computer power supply. Most have at least 5 volts at 10 amps. The newer ATX style power supplies will require a jumper from the green wire on pin 14 or 16 to grouns, a black wire. A tinned jumper wire will fit tightly or some even use apaper clip.

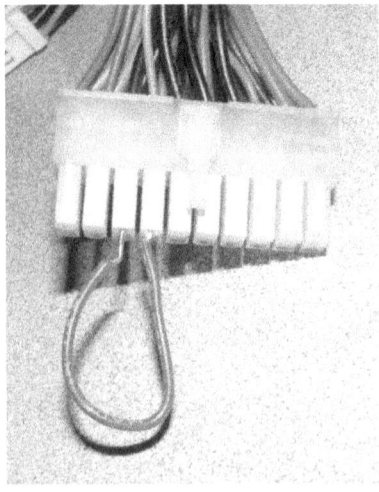

Once you have the power supply jumpered so it is powered on you will have to make a floppy connector to screw terminals or similar adapter.

There is a jumper board that does all of that for you available on eBay. This adapter board has fuses, on off switch, and even a power on light.

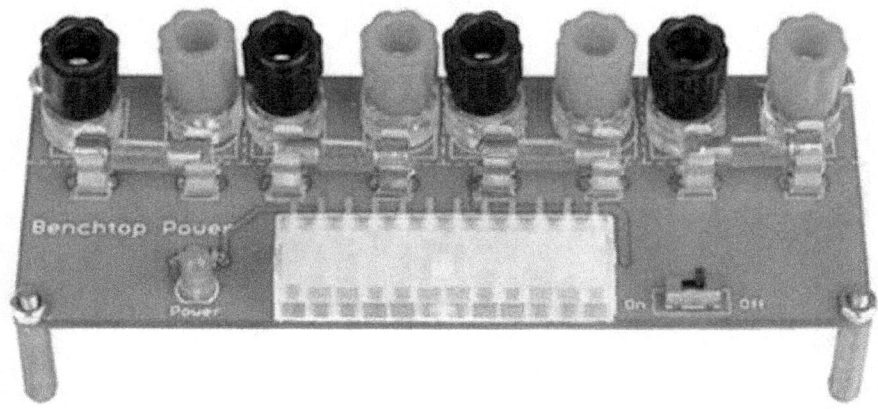

Parts Sources.

I get most of my parts on eBay. However it is good to know who is selling the real thing to avoid getting ripped off. Some people sell rejects that they have "fixed" but usually there are still problems with their products.

The LED strips came from a vendor called electronic.alice. The price shown of $22 is for five meters of the 300 LED/meter strips that are used in the higher density signs. The lower density 150 LED per meter strips are about $18 for five meters of LED's. She also sells the 144 LED/meter strips for about $12 per meter. The 144 LED/meter strips only come in one meter lengths because they are so hard to solder together.

The JST connectors are from epacketshop. The price shown is for 10 sets of connectors. I have used about 50 sets of connectors so far.

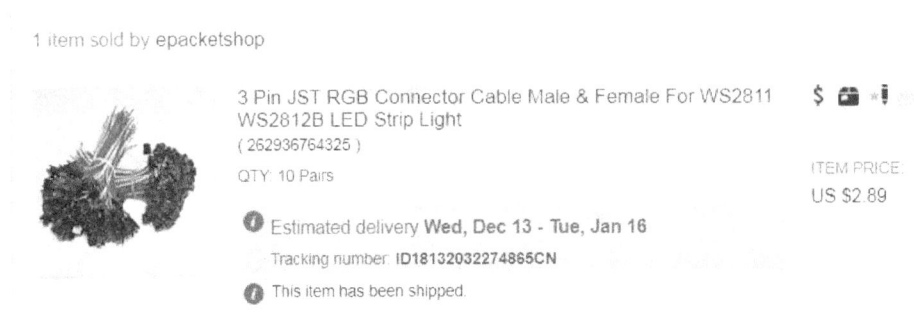

My power supply is only a 12 amp version. I have some of these left over from old LED signs. I would recommend that you get a 20 or 30 amp 5 volt power supply if you are planning on making LED signs. They are only about $20 each.

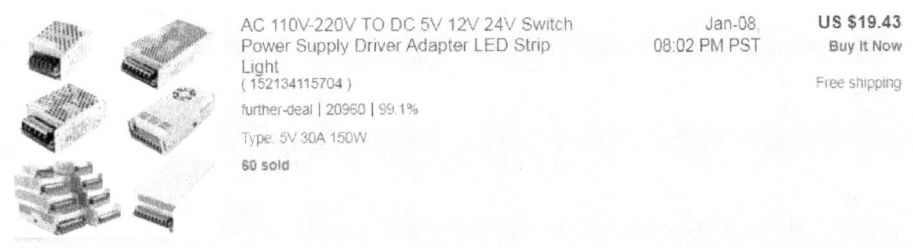

For making things with the WS2811 like a Christmas star or decorating your Christmas tree edirect-us would be a good source. They are about $10 to $15 for 50 of them depending on the vendor.

You can also buy preassembled LED panels from 8x8 to 32x8 in size. An 8x8 (64 LED) panel is about $12 and a 16x16 or 32x8 (256 LED) panel is about $29. So the larger panels are cheaper per LED to buy.

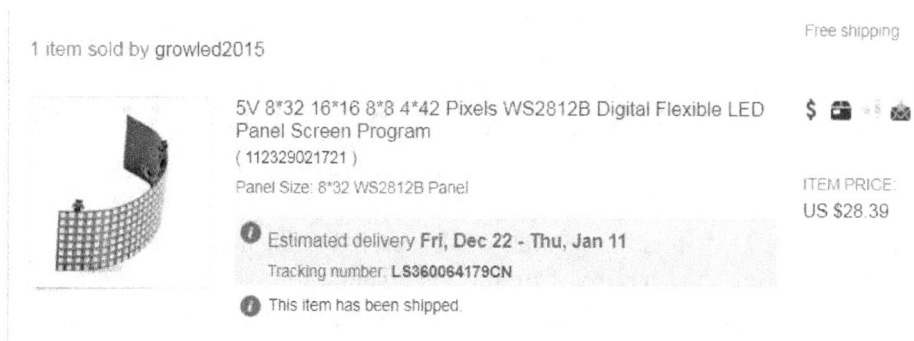

Chapter 2

Running One LED Strip or Array

Running a LED strip is fairly simple. The strips come with three pin connectors soldered on both of their ends. Then there should be an extra connector that you can adapt into an Arduino Uno and a power source. The wires on the extra connector will need to be tinned with solder to make them stiff, or you can solder pins on the ends of the wires. The three wires from the connector go to five volts (red), D6 (green) and ground (white). There should also be two wires that are soldered into the LED strip. They are white for ground and red for five volts power they go to a power supply capable of powering the strip when all LED's are on.

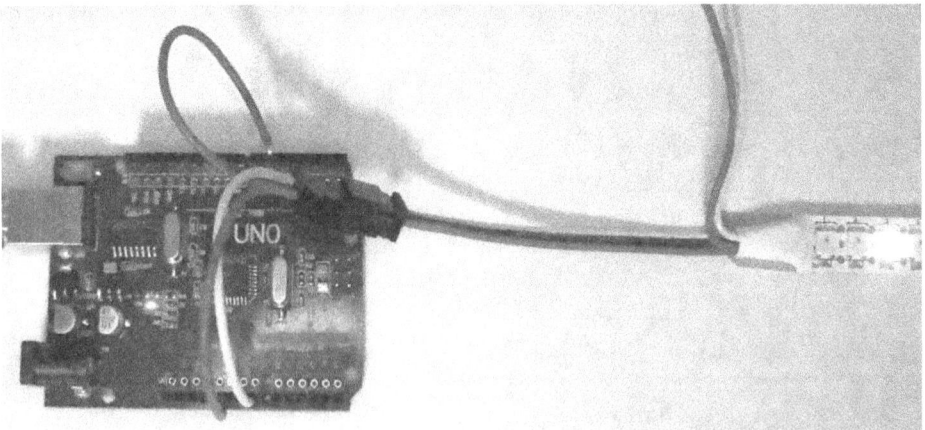

Wire Color Code:
Red is 5 Volts goes to 5V
White is Ground goes to Gnd
Green is Data goes to D6

Then you need to upload a demo program into the Arduino such as the one found later in this chapter and enjoy the effects! Be careful to always connect the five volt power supply before the USB cable to prevent overloading the Arduino's regulator or the USB port of the computer.

8x8 LED Array.

This 8x8 addressable LED array is marked as a CJMCU WS2812 Addressable LED Array. We will interface it with an Arduino Uno.

Someone asked me why I did not include addressable LED arrays in my book "Arduino LED Projects". Well basically back them I did not know about addressable LED's. So recently a friend wanted some LED signs and effects at his wedding reception so I started looking into making things with some addressable LED's. The first thing I bought was a 8x8 LED array. I connected it up to an Arduino and loaded Adafruit's NeoPixel software into the Arduino and got NOTHING!

Then someplace I read that the "Data in" and "Data out" labels are swapped. So, to get it to work, connect the displays "D out" pin to the Arduino D6 pin. It will then come to LIFE!

If you do not provide a separate power supply for the LED array after about a minutes smoke will rise out of your Arduino. The voltage regulator will go up in smoke. Find an 5 volt 1-2 amp AC adapter and use that to power the LED array.

Here is a picture of the 8x8 addressable LED array taken while running the Adafruit demo.

This next picture is the back view showing how to connect the array to the Arduino and to power. In this example the AC adapter is also powering the Arduino. At the bottom of the picture five volts power is coming in on a two pin connector. A four pin jumper is connected to 5 volts, D6, and ground. The fourth jumper pin is not used.

The next diagram shows how the 8x8 array is internally wired. Note for this view the label will be upside down as in the drawing and the "in" jack is labeled "out".

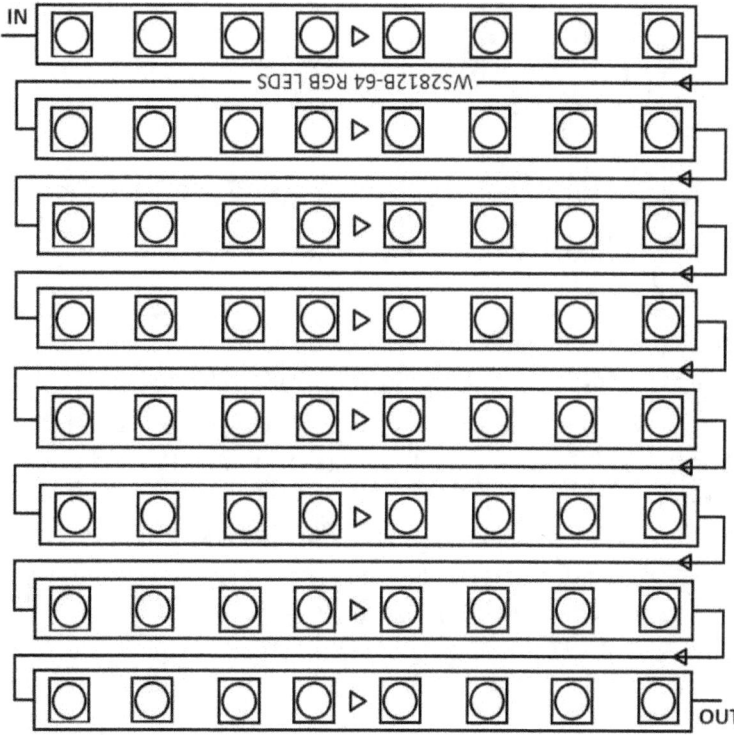

This next program can run one strip or an array. It is set to use D6 but you can easily change that. This program features up to eight shades of brightness on each of the LED's. The brightness and color is specified as three numbers separated by commas such as (4, 4, 4). The program should turn the entire array red, then green, then blue. Then a worm travels down the display. Next there should be a fireworks display of random numbers.

```
// Program for sending data to one WS2812 string
// By Robert Davis II
// Uses Digital Pin 6 change for other pins.
int LEDpin=6;
int MaxLed=256;
int DotBright=64; // sets your brightness

// The delay timing is for an Arduino UNO.
// Sorry digitalWrite is too slow to use.
```

```
void sendBit( uint8_t bits ) {
  PORTD= 0xF0;  // turn on
  PORTD= 0xF0;  // delay
  PORTD= 0xF0;  // delay
  PORTD= 0xF0;  // delay
  PORTD= 0xF0;  // delay (Add more for faster processors)
  PORTD= bits;  // send data
  PORTD= bits;  // delay
  PORTD= bits;  // delay
  PORTD= bits;  // delay
  PORTD= bits;  // delay
  PORTD= 0x00;  // Turn off;
}

void sendColor( int red, int green, int blue ) {
 // Send the bit 8 times down the row, each pixel is 8 bits each for R,G,B
  int mask = 0x01;  // shifting mask to determine bit status
    for (int bit=8; bit>0; bit--){
      if (green & (mask<<bit))sendBit( 0xF0 ); else sendBit( 0x00 ); }
    for (int bit=8; bit>0; bit--){
      if (red & (mask<<bit))sendBit( 0xF0 ); else sendBit( 0x00 ); }
    for (int bit=8; bit>0; bit--){
      if (blue & (mask<<bit))sendBit( 0xF0 ); else sendBit( 0x00 ); }
}

void setup() {
  pinMode (LEDpin, OUTPUT);
  }

void loop() {
  // RED COLOR TEST
  cli();            // No time for interruptions!
  for (int l=0; l<MaxLed; l++){    // number of LED's
    sendColor( l*2, 0, 0);    // Light All red
   }
  delay(500000);
  sei();

  // GREN COLOR TEST
  cli();            // No time for interruptions!
  for (int l=0; l<MaxLed; l++){    // number of LED's
```

```
    sendColor( 0, l*2, 0);      // Light All green
  }
delay(500000);
sei();

// BLUE COLOR TEST
cli();                // No time for interruptions!
for (int l=0; l<MaxLed; l++){    // number of LED's
    sendColor( 0, 0, l*2);       // Light All blue
  }
delay(500000);
sei();

// CLEAR SCREEN
cli();                // No time for interruptions!
for (int l=0; l<MaxLed; l++){    // number of LED
    sendColor( 0, 0, 0);         // clear the screen
}
sei();

// WORM TEST
cli();                // No time for interruptions!
int worm = 1;
for (int cycle=1; cycle<MaxLed; cycle++){
  for (int l=0; l<MaxLed; l++){    // number of LED's
    if (l==worm)   sendColor( DotBright, DotBright, DotBright); // White
    if (l==worm-1) sendColor( DotBright, 0, 0);      // Light red
    if (l==worm-2) sendColor( 0, DotBright, 0);      // Light green
    if (l==worm-3) sendColor( 0, 0, DotBright);      // Light blue
    if (l <worm-3) sendColor( 0, 0, 0);       // Erase old
  }
  worm++;
  delay(20000);
}
delay(20);
sei();

// FIREWORKS
cli();                // No time for interruptions!
for (int cycle=0; cycle<500; cycle++){
    int rano=random(MaxLed);       // Set Random location
```

```
    for (int l=0; l<MaxLed; l++){   // number of bytes to send
      if (l==rano) sendColor( DotBright, DotBright, DotBright);
      else         sendColor( 0, 0, 0);
    }
    delay(20);
  }
 sei();
 return;
}
```

15x16 Array

I purchased two five meter, almost 200 inches long WS2812 addressable LED array strips. Then I separated them into ten strips of 15 LED's each about 20 inches long. Why 15 LED's? Because that is where they are soldered together at the factory. Then they were stuck on "yard sign" material for a backing. The result is a 20 inch by 20 inch sized 15 by 16 LED array.

This next picture shows where the strips are soldered together at the factory. The solder junction is in the center of the picture. This is the easiest point to separate the strips into smaller pieces. Use a soldering iron with a long tip and heat all three connections. At the same time bend the strip in a U shape and it will then separate.

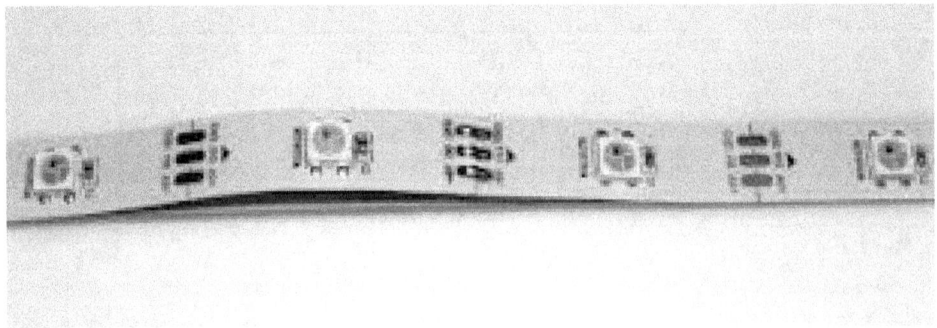

This picture shows the 5 volt 12 amp power supply powering the array. The 15x16 array takes about four to five amps to power it. At first I wired the array in the zigzag arrangement where each strip goes in the opposite direction. However that makes creating text a bit more difficult.

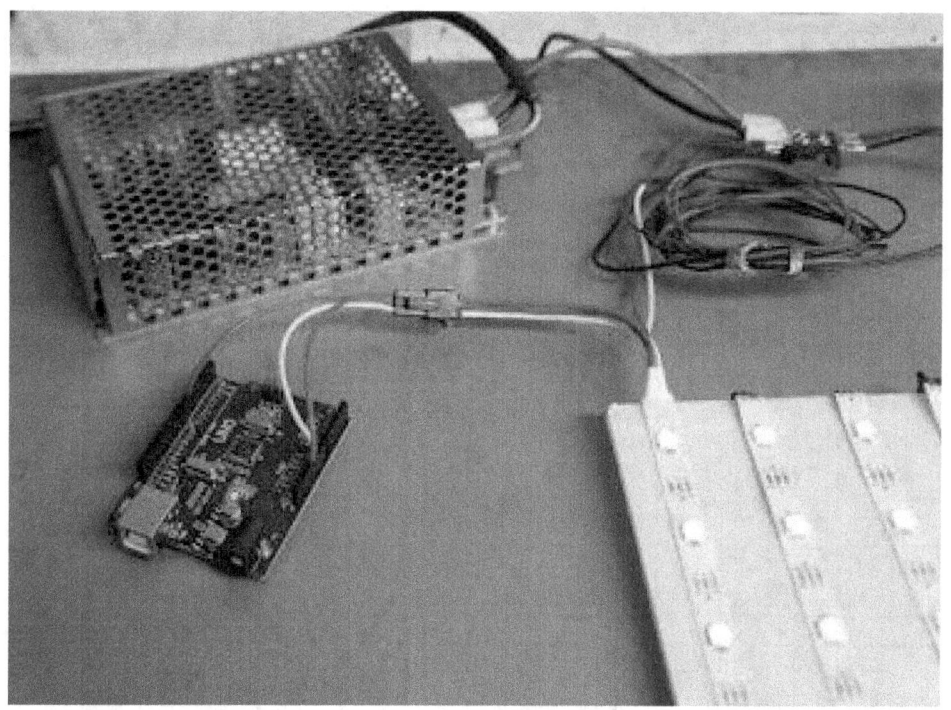

Wiring this array so each segment starts at the top will require 48 two foot long jumpers to loop the power and data lines from the bottom back up to the top 16 times. The wiring is all hidden on the back side of the array. Actually you only need to run the data lines from the bottom to the top. The power and ground lines can be connected together across the top and bottom.

Make sure all the arrows on the LED strip segments are pointed in the same direction. Also you should connect power to the tail end connector as there is a lot of power going through the two now rearranged LED strips.

This next schematic diagram shows how the LED strips are connected together to get each segment to work from top to bottom. It is a good idea to also connect power at the bottom right corner.

There is a timing limit of about 1000 LED's using this method. Sending the data out serially takes time. You have to send three eight bit bytes to each LED. that is 24 times 100 or 24000 bytes of data to send. By crunching the timing you might be able to get up to 2000 LED's to work. But if you could run several display strips at the same time you can go beyond that limit to 1000 x 8 or 8000 LED's or even more. We will look into using that arrangement to drastically increase the amount of LED's in the next chapter.

8x32 Array

There is a flexible 8x32 addressable LED array that is available on eBay. I was hoping that it was wired up the same way as the 8x8 array was. But it is wired in the zigzag manner instead. Basically each strip is rotated so that it goes in the opposite direction of the previous one. There are also some separate power and ground connections that are located in the middle of the array. This zigzag wiring makes doing things like characters or text a bit tricky to do. However by reversing the order of each column, displaying text can be done.

This picture shows how to connect a 8x32 Array up to an Arduino Uno. It connects just like the single LED strip does. Red is 5V, Green is D6, and white is ground. On the power connector black is ground and red is five volts.

The back of the LED array has input three pin, output three pin, and two wire power connections.

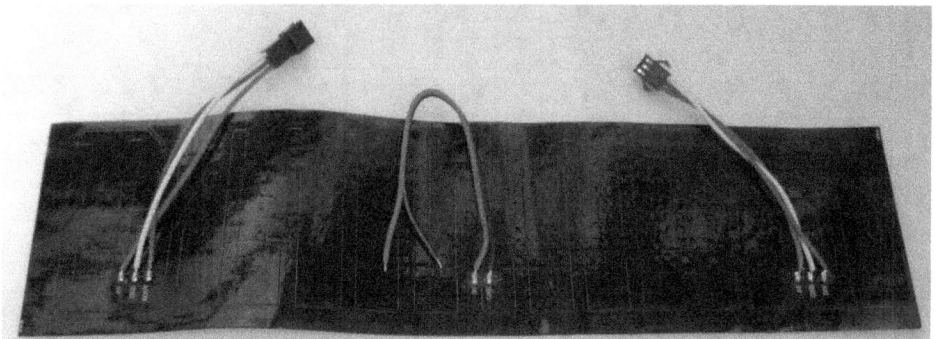

The next schematic diagram shows how the 8x32 flexible LED array is wired up internally with the internal zigzag format. As you can see the data arrows are reversed in every other strip. I did not draw the fact that the power is connected up to it in the center of the strip. Mine came with the three pin input and output connectors and the power connector already soldered on to the array.

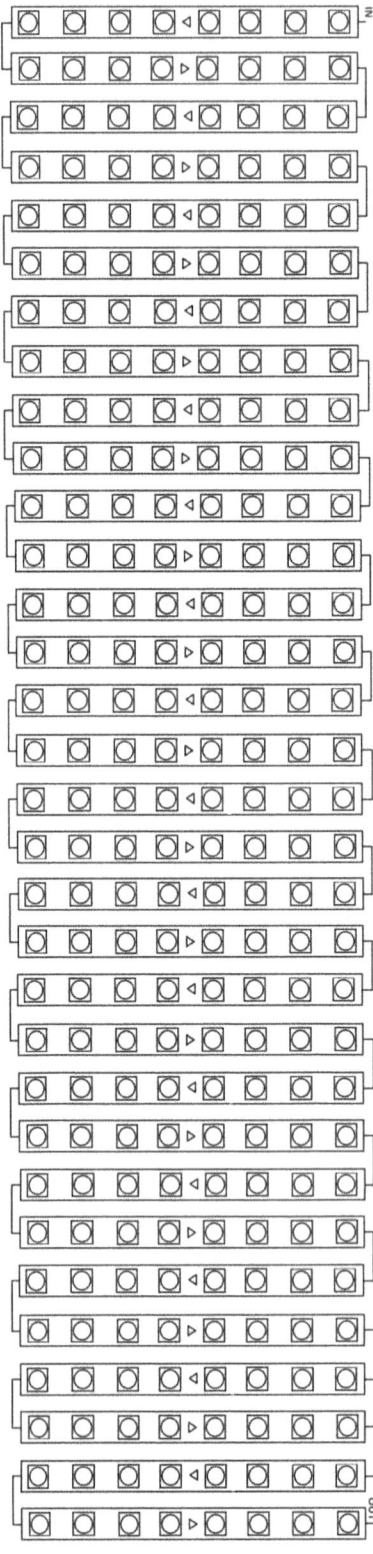

This next program displays text on a eight bit zigzag wired LED display. Note the lack of a loop when displaying text column bits to make it faster and that also makes it easier to see that the column bit masking reverses with each column.

```
// Program for sending data to one WS2812 string
// ZIGZAG text displayed
// By Robert Davis II
// Uses Digital Pin 6 change for other pins.
int LEDpin=6;
int MaxLed=256;  // how many LED's
char PixelRow;
int textc=128;  // brightness of text

// Set default color for letters
int red=1; int green=1; int blue=1;

// The delay timing is for an Arduino UNO.
// Sorry digitalWrite is too slow to use.
void sendBit( uint8_t bits ) {
  PORTD= 0xF0;  // turn on
  PORTD= 0xF0;  // delay
  PORTD= 0xF0;  // delay
  PORTD= 0xF0;  // delay
  PORTD= 0xF0;  // delay (Add more for faster processors)
  PORTD= bits;  // send data
  PORTD= bits;  // delay
  PORTD= bits;  // delay
  PORTD= bits;  // delay
  PORTD= bits;  // delay
  PORTD= 0x00;  // Turn off;
}

void sendColor( uint8_t row ) {
// Send the bit 8 times down the row, each pixel is 8 bits each for R,G,B
 int mask = 0x01;  // shifting mask to determine bit status
   for (int bit=8; bit>0; bit--){
    if (green & (mask<<bit))sendBit( row ); else sendBit( 0x00 ); }
   for (int bit=8; bit>0; bit--){
    if (red & (mask<<bit))sendBit( row ); else sendBit( 0x00 ); }
   for (int bit=8; bit>0; bit--){
```

```
    if (blue & (mask<<bit))sendBit( row ); else sendBit( 0x00 ); }
}

char text1[]="ARDUINO POWERED WS2812 LED SIGN                    ";

// This font from http://sunge.awardspace.com/glcd-sd/node4.html
byte font[][6] = {
0x00,0x00,0x00,0x00,0x00,0x00, // ascii 32
0x00,0x00,0xfa,0x00,0x00,0x00, // !
0x00,0xe0,0x00,0xe0,0x00,0x00, // "
0x28,0xfe,0x28,0xfe,0x28,0x00, // #
0x24,0x54,0xfe,0x54,0x48,0x00, // $
0xc4,0xc8,0x10,0x26,0x46,0x00, // %
0x6c,0x92,0xaa,0x44,0x0a,0x00, // &
0x00,0xa0,0xc0,0x00,0x00,0x00, // '
0x00,0x38,0x44,0x82,0x00,0x00, // (
0x00,0x82,0x44,0x38,0x00,0x00, // )
0x10,0x54,0x38,0x54,0x10,0x00, // *
0x10,0x10,0x7c,0x10,0x10,0x00, // +
0x00,0x0a,0x0c,0x00,0x00,0x00, // ,
0x10,0x10,0x10,0x10,0x10,0x00, // -
0x00,0x06,0x06,0x00,0x00,0x00, // .
0x04,0x08,0x10,0x20,0x40,0x00, // /
0x7c,0x8a,0x92,0xa2,0x7c,0x00, // 0
0x00,0x42,0xfe,0x02,0x00,0x00, // 1
0x42,0x86,0x8a,0x92,0x62,0x00, // 2
0x84,0x82,0xa2,0xd2,0x8c,0x00, // 3
0x18,0x28,0x48,0xfe,0x08,0x00, // 4
0xe4,0xa2,0xa2,0xa2,0x9c,0x00, // 5
0x3c,0x52,0x92,0x92,0x0c,0x00, // 6
0x80,0x8e,0x90,0xa0,0xc0,0x00, // 7
0x6c,0x92,0x92,0x92,0x6c,0x00, // 8
0x60,0x92,0x92,0x94,0x78,0x00, // 9
0x00,0x6c,0x6c,0x00,0x00,0x00, // :
0x00,0x6a,0x6c,0x00,0x00,0x00, // ;
0x00,0x10,0x28,0x44,0x82,0x00, // <
0x28,0x28,0x28,0x28,0x28,0x00, // =
0x82,0x44,0x28,0x10,0x00,0x00, // >
0x40,0x80,0x8a,0x90,0x60,0x00, // ?
0x4c,0x92,0x9e,0x82,0x7c,0x00, // @
0x7e,0x90,0x90,0x90,0x7e,0x00, // A
```

```
0xfe,0x92,0x92,0x92,0x6c,0x00, // B
0x7c,0x82,0x82,0x82,0x44,0x00, // C
0xfe,0x82,0x82,0x44,0x38,0x00, // D
0xfe,0x92,0x92,0x92,0x82,0x00, // E
0xfe,0x90,0x90,0x80,0x80,0x00, // F
0x7c,0x82,0x82,0x8a,0x4c,0x00, // G
0xfe,0x10,0x10,0x10,0xfe,0x00, // H
0x00,0x82,0xfe,0x82,0x00,0x00, // I
0x04,0x02,0x82,0xfc,0x80,0x00, // J
0xfe,0x10,0x28,0x44,0x82,0x00, // K
0xfe,0x02,0x02,0x02,0x02,0x00, // L
0xfe,0x40,0x20,0x40,0xfe,0x00, // M
0xfe,0x20,0x10,0x08,0xfe,0x00, // N
0x7c,0x82,0x82,0x82,0x7c,0x00, // O
0xfe,0x90,0x90,0x90,0x60,0x00, // P
0x7c,0x82,0x8a,0x84,0x7a,0x00, // Q
0xfe,0x90,0x98,0x94,0x62,0x00, // R
0x62,0x92,0x92,0x92,0x8c,0x00, // S
0x80,0x80,0xfe,0x80,0x80,0x00, // T
0xfc,0x02,0x02,0x02,0xfc,0x00, // U
0xf8,0x04,0x02,0x04,0xf8,0x00, // V
0xfe,0x04,0x18,0x04,0xfe,0x00, // W
0xc6,0x28,0x10,0x28,0xc6,0x00, // X
0xc0,0x20,0x1e,0x20,0xc0,0x00, // Y
0x86,0x8a,0x92,0xa2,0xc2,0x00, // Z
0x00,0x00,0xfe,0x82,0x82,0x00, // [
0x40,0x20,0x10,0x08,0x04,0x00, // slash
0x82,0x82,0xfe,0x00,0x00,0x00, // ]
0x20,0x40,0x80,0x40,0x20,0x00, // ^
0x02,0x02,0x02,0x02,0x02,0x00, // _
0x00,0x80,0x40,0x20,0x00,0x00, // `
0x00,0x10,0x6c,0x82,0x00,0x00, // {
0x00,0x00,0xfe,0x00,0x00,0x00, // |
0x00,0x82,0x6c,0x10,0x00,0x00, // }
0x10,0x10,0x54,0x38,0x10,0x00, // ~
0x10,0x38,0x54,0x10,0x10,0x00, // •
};

void setup() {
 pinMode (LEDpin, OUTPUT);
 }
```

```
void loop() {
 // RED COLOR TEST
 red=1; green=0; blue=0;
 cli();              // No time for interruptions!
 for (int l=0; l<MaxLed/4; l++){    // number of LED's
  red=l;
  sendColor( 0xF0);      // Light All red
  }
 // GREEN COLOR TEST
 red=0; green=1; blue=0;
 for (int l=0; l<MaxLed/4; l++){    // number of LED's
  green=l;
  sendColor( 0xF0);      // Light All green
  }
 // BLUE COLOR TEST
 red=0; green=0; blue=1;
 for (int l=0; l<MaxLed/4; l++){    // number of LED's
  blue=l;
  sendColor( 0xF0);      // Light All blue
  }
 // WHITE COLOR TEST
 for (int l=0; l<MaxLed/4; l++){    // number of LED's
  red=l; green=l; blue=l;
  sendColor( 0xF0);      // Light All white
  }
 sei();
 delay(2000);

 //TEXT with ZIGZAG layout - bit order reverses for each column.
 cli();              // No time for interruptions!
 for (int l=0; l<30; l++){    // 30 is number of characters to send
   switch (l){         // case is faster than if's
   case 0: red=textc; green=0; blue=0; break;
   case 1: red=0; green=textc; blue=0; break;
   case 2: red=0; green=0; blue=textc; break;
   case 3: red=textc; green=0; blue=0; break;
   case 4: red=0; green=textc; blue=0; break;
   case 5: red=0; green=0; blue=textc; break;
   }
  for (int r=0; r<6; r++) {
```

```
      PixelRow=(font[text1[l]-32][r]); // -32 is to get to correct ascii
      if (r & 0x01){     // invert odd rows
      // Using a loop and shifting is too slow with the character lookups.
      if (PixelRow & 0x01 ) sendColor (0xF0); else sendColor (0x00);
      if (PixelRow & 0x02 ) sendColor (0xF0); else sendColor (0x00);
      if (PixelRow & 0x04 ) sendColor (0xF0); else sendColor (0x00);
      if (PixelRow & 0x08 ) sendColor (0xF0); else sendColor (0x00);
      if (PixelRow & 0x10 ) sendColor (0xF0); else sendColor (0x00);
      if (PixelRow & 0x20 ) sendColor (0xF0); else sendColor (0x00);
      if (PixelRow & 0x40 ) sendColor (0xF0); else sendColor (0x00);
      if (PixelRow & 0x80 ) sendColor (0xF0); else sendColor (0x00);
      }
      else{
      if (PixelRow & 0x80 ) sendColor (0xF0); else sendColor (0x00);
      if (PixelRow & 0x40 ) sendColor (0xF0); else sendColor (0x00);
      if (PixelRow & 0x20 ) sendColor (0xF0); else sendColor (0x00);
      if (PixelRow & 0x10 ) sendColor (0xF0); else sendColor (0x00);
      if (PixelRow & 0x08 ) sendColor (0xF0); else sendColor (0x00);
      if (PixelRow & 0x04 ) sendColor (0xF0); else sendColor (0x00);
      if (PixelRow & 0x02 ) sendColor (0xF0); else sendColor (0x00);
      if (PixelRow & 0x01 ) sendColor (0xF0); else sendColor (0x00);
      }
    }
  }
sei();
delay(4000);              // Wait to display results

// FIREWORKS
red=textc; green=textc; blue=textc;
cli();              // No time for interruptions!
for (int cycle=0; cycle<500; cycle++){
  int rano=random(MaxLed);       // Set Random location
  for (int l=0; l<MaxLed; l++){   // number of bytes to send
    if (l==rano) sendColor( 0xF0); else sendColor( 0x00);
  }
  delay(20);
}
sei();
return;
}
```

Chapter 3

Interfacing to Eight LED Strips

30 LED per Meter Sign

I found out that there is a supposed 500 LED limit on running continuous LED arrays from one output of the Arduino. However you can run over 1000 LED's that way. Then I discovered an alternate setup that can run over 8000 addressable LED's from an Arduino Uno. The alternate method uses the parallel output mode to run eight strings of LED's at the same time.

This picture shows the wiring, this is much easier than any other method of running this many addressable LED strings. Since I only had four strings of 150 LED's at that time, they were split in half to make 8 strings of 75 LED's each. This is an early picture of the wiring back when it first started working. I had to cobble some of the connections. Later on I purchased a pile of the three pin connectors on eBay and expanded to using all of the eight data lines.

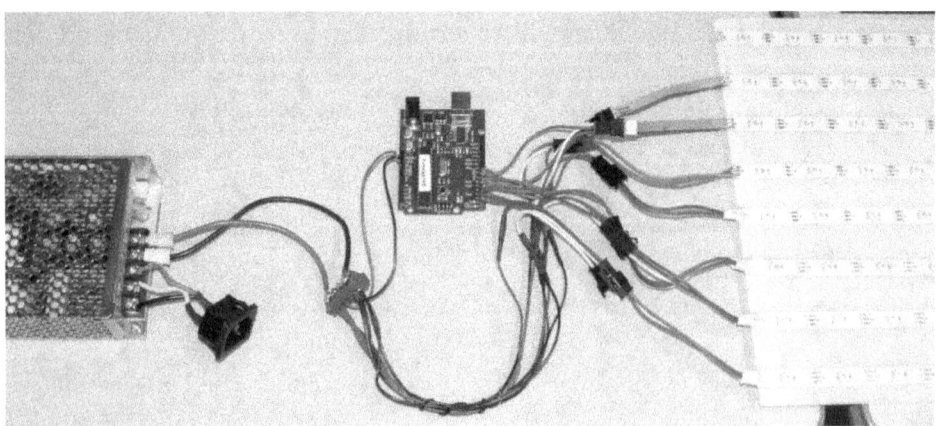

To connect the Arduino to the LED sign you will need to make a wiring harness. Basically the eight data lines connect to D0 to D7 and the power and ground wires connect to the power supply. Here is a picture of the wiring harness.

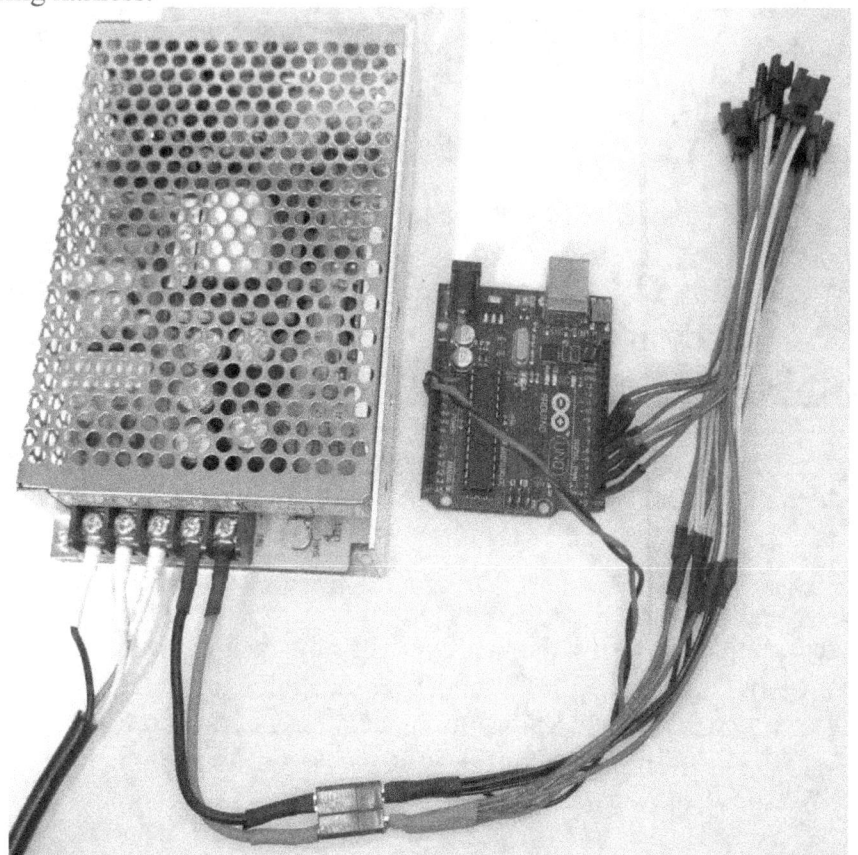

Coming up next there are two pictures of my test program running on my first big LED sign. At that time it was only up to 600 LED's configured as eight rows of 75 LED's each. I had changed from displaying an array of patterns into displaying some text from a character generator. I have also improved the code to support much longer strings and added a complete character generator.

The software also supported only up to eight colors at that time. Later on it was expanded to support 256 shades of each of three colors. These pictures do not show you how bright the colors are because the camera automatically turns down the brightness.

The first picture is bright red and the second picture is bright green. I also did the sign in bright blue as seen on my YouTube video.

The sign consisted of three interconnecting panels. The panel seams are visible to the left of the A and to the left of the O in the first picture and between the L and E and between the I and G in the second picture.

This next picture is a schematic diagram of how to wire up the panels. The schematic only shows the first ten LED's that are on the strips. A 40 inch panel will have 30 LEDs or 60 LED's depending on the number of LED's per inch. A 48 inch (four foot) panel does not line up with the seams in the LED strips, but you can make custom LED strip lengths to work with 48 inch wide panels.

You can also choose to do a 60 inch (5 foot) long panel size as it will also line up with the factory LED segments. My biggest LED sign panel is five foot long. It was made out of a four foot by eight foot sign board that a friend gave me. However most sign board material is purchased in smaller sizes.

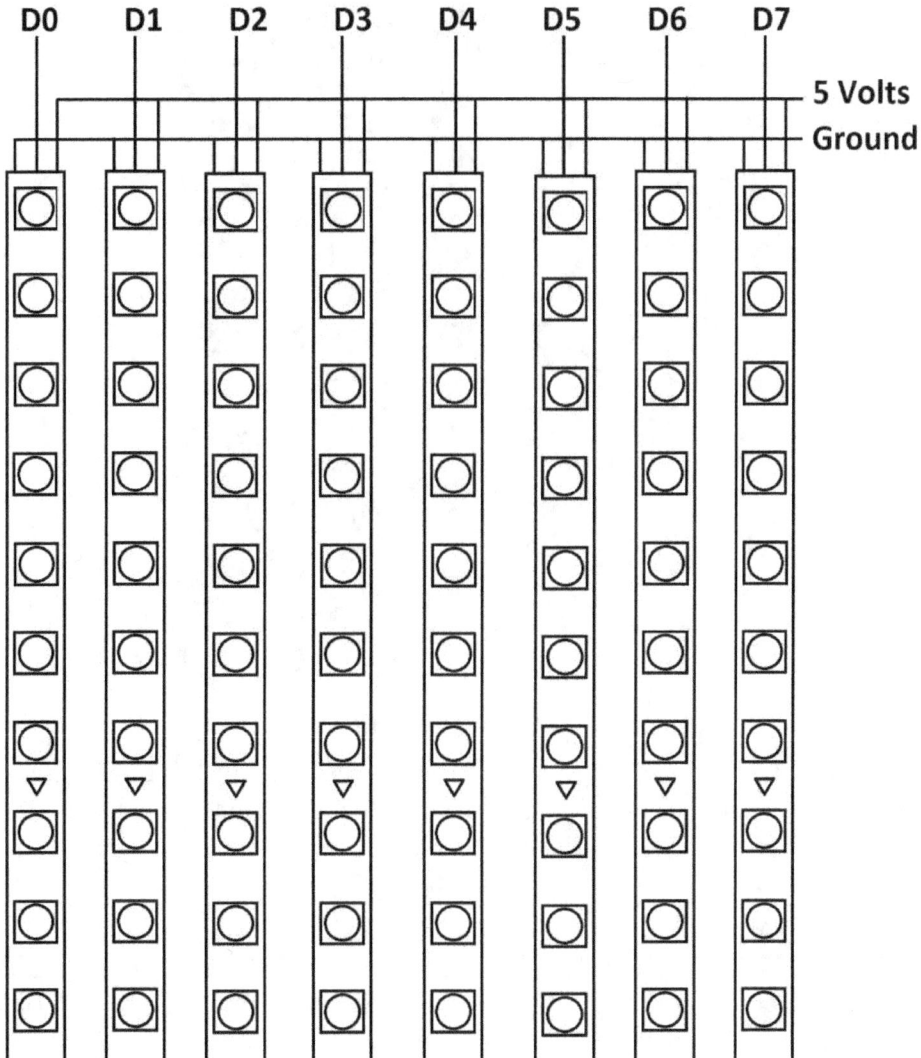

The sign also breaks down into multiple sections that are each 40 inches or 30 or 60 LED's long. That is done with eight 3 pin plugs that are identical to the connectors that comes on the LED strips. The wiring was folded over at the end of the panel and held in place on the back side using 1 inch wide strips of sign material. Using a good clear stretching glue gives a much cleaner appearance. This next picture shows what the interconnects look like.

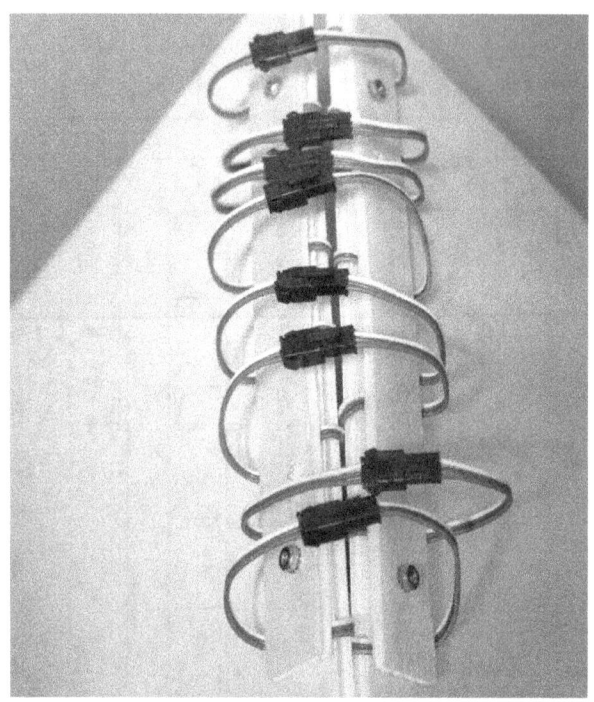

If you have 30 LED's per meter and 40 inch by 10 inch sign board then this is what you will get. That is 30 LED's times eight or 240 LED's per board. I have three of these panels that can be interconnected for a 90 x 8 or 720 LED sign that is about 10 feet long.

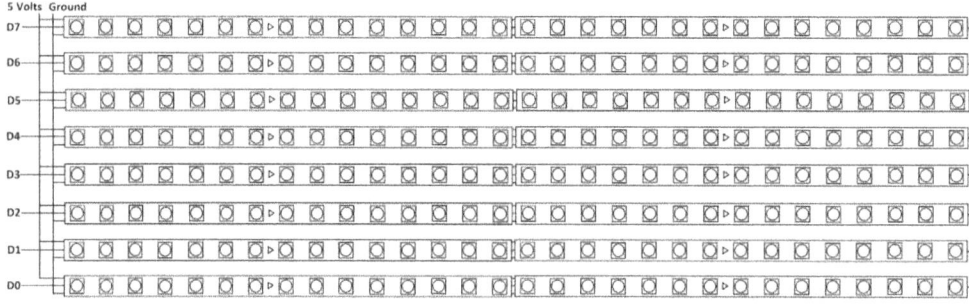

60 LED per Meter Sign

I also have a higher density addressable LED sign. It is five feet by one foot in size. With the 60 LED per meter strips that works out to having 90 LED's in each of the five foot long strips. The sign has 16 of these 5 foot long strips. The output of the top row for goes to the input of the ninth row. The output of the second row goes to the input of the tenth row, etc.

So the sign appears electrically as eight rows of 180 LED's each. That is 1440 LED's total.

Here are two pictures showing the front and back sides of the 60 LED per meter LED sign. It takes eight - three conductor five foot long jumper cables to go from the top eight rows to the bottom eight rows as can be seen in the picture of the back view.

The front view shows the 1440 addressable LED's as 16 rows of 90 LED's.

The back view shows the eight five foot long male to female jumper wires.

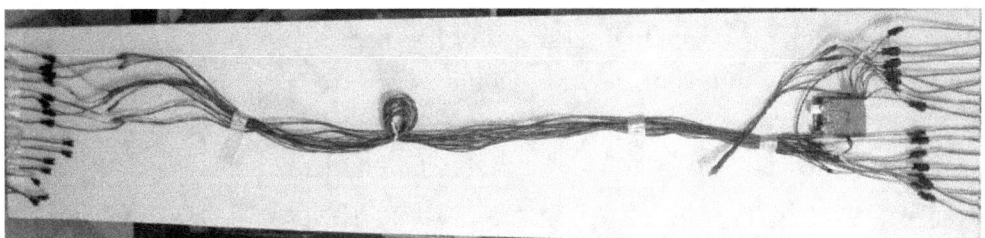

This picture shows the 60 LED per meter sign displaying some sample text in multiple colors.

The next schematic shows how D0b to D7b are then ends of D0 to D7 looped back from the top right side to the bottom left side. It shows 30 LED's per strip but the sign actually uses 90 LED's per strip. That is 60 inches of 60 LED per meter (40 inches) LED's spaced every 7/8 inches.

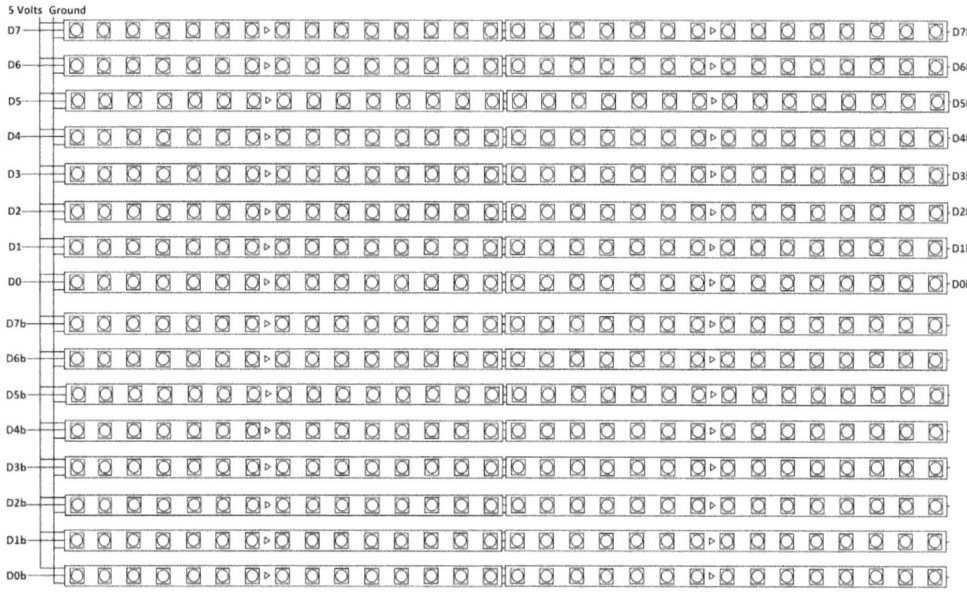

144 LED per Meter Sign

I also built a third sign that has the 144 LED per meter or 1/4 inch LED spacing. This sign is made with the third common LED strip spacing. I do not know why they do not seem to sell a 90 LED per Meter LED strip. The first tests of the 144 LED per meter sign indicated that it had lots of bad connections. The issue was likely caused by the LED strips being shipped while wrapped on a spool that has too small of a diameter.

The LED's are so close together on the 144 LED per meter strips that joining the strips together at the factory requires a larger space between the strips as can be seen in this next picture. Because of that limitation I decided to make my sign only 20 inches or one strip wide. With some practice I am sure that the spacing for soldering could be reduced.

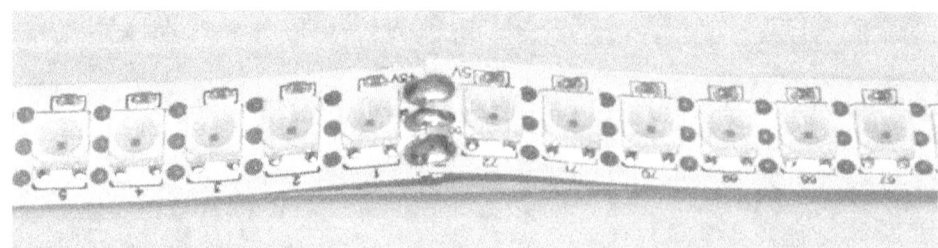

My 144 LED per meter had lots of problems. In this picture the middle row quits about 3/4 the way across the sign. One of the pixels in the "U" is red but that is hard to tell in the picture.

Some other issues were that the bottom right corner LED was out. The third row down is a different shade of blue but you cannot see that in the picture. Most of the problems were fixed by re-soldering the faulty LED's. Also re-solder one LED to the left of a faulty LED as sometimes the output of that one is the issue.

When I connected the remainder of the sign the results were even worse! Two rows stopped in the middle and five rows did not even work at all! I purchased some more strips to replace the faulty strips and re-soldered more bad connections.

This would be a good time to discuss troubleshooting the LED strips. Sometimes I used some 30 gauge wire wrap wire to jump across bad connections. The next picture shows how the strips are wired internally. You can use a meter with a beeper to check for the bad connections with the sign powered off. If one LED does not work check the power and ground connections. If a strip of LED's stops working check the Data out to data in connections. Sometimes pressing with the meter probes will make the problem go away temporarily.

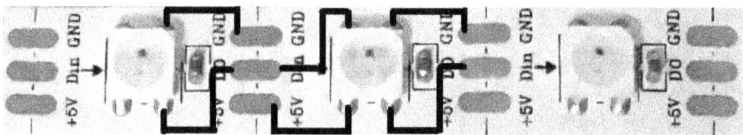

Another issue is soldering in the replacement LED's when you have bad LED's. Never try to solder to the LED's themselves. They are too easily damaged. Cut out the bad section of the strip in the center of the pads. Do the same for the replacement section. Then tin all of the pads. Then insert the replacement strip into the sign and solder the ends together. The ends should not overlap much, only a tiny bit.

The schematic diagram for the 144 LED per meter sign is the same as for the 60 LED per meter sign. However the sign I made was only 20 inches wide by 10 inches tall so it was only 72 LED's across. Instead of using jumpers across the back to connect the top eight rows to the bottom eight rows I extended the length of the connectors on the right side by about 14 inches so they reached the other side.

Multicolor Text Sign Program

Here is the code to make any of the eight bit parallel LED signs work:

```
// Version for sending data to 8 Parallel WS2812 strings
// by: BOB Davis
// Removed Assembler and simplified the code

// PORTD is Digital Pins 0-7 on the Uno change for other boards.
#define PIXEL_PORT  PORTD  // Port the pixels are connected to
#define PIXEL_DDR   DDRD   // Port the pixels are connected to
int textb=128;  // brightness of text
```

```
int maxled=180;

// Send the next set of 8 WS2812B encoded bits to the 8 pins.
// The delay timing is for an Arduino UNO.
void sendBitX8( uint8_t bits ) {
  PORTD= 0xFF;  // turn on
  PORTD= 0xFF;  // delay
  PORTD= 0xFF;  // delay
  PORTD= 0xFF;  // delay
  PORTD= 0xFF;  // delay (add more for faster processors)
  PORTD= bits;  // send data
  PORTD= bits;  // delay
  PORTD= bits;  // delay
  PORTD= bits;  // delay
  PORTD= bits;  // delay
  PORTD= 0x00;  // Turn off;
}

// Set default color for letters
int red=1; int green=1; int blue=1;

void sendPixelRow( uint8_t row ) {
  // Send the bit 8 times down every row, each pixel is 8 bits/color
  int mask = 0x01; // shifting mask to determine bit status
    for (int bit=8; bit>0; bit--){
      if (green & (mask<<bit))sendBitX8( row ); else sendBitX8( 0x00 ); }
    for (int bit=8; bit>0; bit--){
      if (red & (mask<<bit))sendBitX8( row ); else sendBitX8( 0x00 ); }
    for (int bit=8; bit>0; bit--){
      if (blue & (mask<<bit))sendBitX8( row ); else sendBitX8( 0x00 ); }
}

char text[]="ARDUINO ADXABLE LED PROJECTS   ";

// This font from http://sunge.awardspace.com/glcd-sd/node4.html
byte font[][6] = {
0x00,0x00,0x00,0x00,0x00,0x00, // ascii 32
0x00,0x00,0xfa,0x00,0x00,0x00, // !
0x00,0xe0,0x00,0xe0,0x00,0x00, // "
0x28,0xfe,0x28,0xfe,0x28,0x00, // #
0x24,0x54,0xfe,0x54,0x48,0x00, // $
```

```
0xc4,0xc8,0x10,0x26,0x46,0x00, // %
0x6c,0x92,0xaa,0x44,0x0a,0x00, // &
0x00,0xa0,0xc0,0x00,0x00,0x00, // '
0x00,0x38,0x44,0x82,0x00,0x00, // (
0x00,0x82,0x44,0x38,0x00,0x00, // )
0x10,0x54,0x38,0x54,0x10,0x00, // *
0x10,0x10,0x7c,0x10,0x10,0x00, // +
0x00,0x0a,0x0c,0x00,0x00,0x00, // ,
0x10,0x10,0x10,0x10,0x10,0x00, // -
0x00,0x06,0x06,0x00,0x00,0x00, // .
0x04,0x08,0x10,0x20,0x40,0x00, // /
0x7c,0x8a,0x92,0xa2,0x7c,0x00, // 0
0x00,0x42,0xfe,0x02,0x00,0x00, // 1
0x42,0x86,0x8a,0x92,0x62,0x00, // 2
0x84,0x82,0xa2,0xd2,0x8c,0x00, // 3
0x18,0x28,0x48,0xfe,0x08,0x00, // 4
0xe4,0xa2,0xa2,0xa2,0x9c,0x00, // 5
0x3c,0x52,0x92,0x92,0x0c,0x00, // 6
0x80,0x8e,0x90,0xa0,0xc0,0x00, // 7
0x6c,0x92,0x92,0x92,0x6c,0x00, // 8
0x60,0x92,0x92,0x94,0x78,0x00, // 9
0x00,0x6c,0x6c,0x00,0x00,0x00, // :
0x00,0x6a,0x6c,0x00,0x00,0x00, // ;
0x00,0x10,0x28,0x44,0x82,0x00, // <
0x28,0x28,0x28,0x28,0x28,0x00, // =
0x82,0x44,0x28,0x10,0x00,0x00, // >
0x40,0x80,0x8a,0x90,0x60,0x00, // ?
0x4c,0x92,0x9e,0x82,0x7c,0x00, // @
0x7e,0x90,0x90,0x90,0x7e,0x00, // A
0xfe,0x92,0x92,0x92,0x6c,0x00, // B
0x7c,0x82,0x82,0x82,0x44,0x00, // C
0xfe,0x82,0x82,0x44,0x38,0x00, // D
0xfe,0x92,0x92,0x92,0x82,0x00, // E
0xfe,0x90,0x90,0x80,0x80,0x00, // F
0x7c,0x82,0x82,0x8a,0x4c,0x00, // G
0xfe,0x10,0x10,0x10,0xfe,0x00, // H
0x00,0x82,0xfe,0x82,0x00,0x00, // I
0x04,0x02,0x82,0xfc,0x80,0x00, // J
0xfe,0x10,0x28,0x44,0x82,0x00, // K
0xfe,0x02,0x02,0x02,0x02,0x00, // L
0xfe,0x40,0x20,0x40,0xfe,0x00, // M
```

```
0xfe,0x20,0x10,0x08,0xfe,0x00, // N
0x7c,0x82,0x82,0x82,0x7c,0x00, // O
0xfe,0x90,0x90,0x90,0x60,0x00, // P
0x7c,0x82,0x8a,0x84,0x7a,0x00, // Q
0xfe,0x90,0x98,0x94,0x62,0x00, // R
0x62,0x92,0x92,0x92,0x8c,0x00, // S
0x80,0x80,0xfe,0x80,0x80,0x00, // T
0xfc,0x02,0x02,0x02,0xfc,0x00, // U
0xf8,0x04,0x02,0x04,0xf8,0x00, // V
0xfe,0x04,0x18,0x04,0xfe,0x00, // W
0xc6,0x28,0x10,0x28,0xc6,0x00, // X
0xc0,0x20,0x1e,0x20,0xc0,0x00, // Y
0x86,0x8a,0x92,0xa2,0xc2,0x00, // Z
};

void setup() {
  PIXEL_DDR = 0xff;     // Set all row pins to output
}
void loop() {
// 256 SHADE COLOR TEST
  cli();                // No time for interruptions!
  for (int l=0; l<(maxled/4); l++){ // max number lines to send
    red=l+2; green=0; blue=0;
    sendPixelRow(0xFF);      // Light red
  }
  for (int l=0; l<(maxled/4); l++){ // max number lines to send
    red=0; green=l+2; blue=0;
    sendPixelRow(0xFF);      // Light green
  }
  for (int l=0; l<(maxled/4); l++){ // max number lines to send
    red=0; green=0; blue=l+2;
    sendPixelRow(0xFF);      // Light blue
  }
  for (int l=0; l<(maxled/4); l++){ // max number lines to send
    red=l+2; green=l+2; blue=l+2;
    sendPixelRow(0xFF);      // Light white
  }
  sei();
  delay(4000);          // Wait to display results

  cli();                // No time for interruptions!
```

```
for (int l=0; l<30; l++){     // 30 is number of characters to send
  switch (l){              // case is faster than if's
    case 0: red=textb; green=0; blue=0; break; // red
    case 1: red=0; green=textb; blue=0; break;  // green
    case 2: red=0; green=0; blue=textb; break;  // blue
    case 3: red=textb; green=textb; blue=0; break; //yellow
    case 4: red=0; green=textb; blue=textb; break; // cyan
    case 5: red=textb; green=0; blue=textb; break; // purple
    case 6: red=textb; green=0; blue=0; break; // red
    case 7: red=0; green=textb; blue=0; break;  // green
    case 8: red=0; green=0; blue=textb; break;  // blue
    case 9: red=textb; green=textb; blue=0; break; //yellow
    case 10: red=0; green=textb; blue=textb; break; // cyan
    case 11: red=textb; green=0; blue=textb; break; // purple
    case 12: red=textb; green=0; blue=0; break; // red
    case 13: red=0; green=textb; blue=0; break;  // green
    case 14: red=0; green=0; blue=textb; break;  // blue
    case 15: red=textb; green=textb; blue=0; break; //yellow
    case 16: red=0; green=textb; blue=textb; break; // cyan
    case 17: red=textb; green=0; blue=textb; break; // purple
    case 18: red=textb; green=0; blue=0; break; // red
    case 19: red=0; green=textb; blue=0; break;  // green
    case 20: red=0; green=0; blue=textb; break;  // blue
    case 21: red=textb; green=textb; blue=0; break; //yellow
    case 22: red=0; green=textb; blue=textb; break; // cyan
    case 23: red=textb; green=0; blue=textb; break; // purple
    case 24: red=textb; green=0; blue=0; break; // red
    case 25: red=0; green=textb; blue=0; break;  // green
    case 26: red=0; green=0; blue=textb; break;  // blue
    case 27: red=textb; green=textb; blue=0; break; //yellow
    case 28: red=0; green=textb; blue=textb; break; // cyan
    case 29: red=textb; green=0; blue=textb; break; // purple
  }
  // Using a loop is too slow with the character lookups.
  sendPixelRow(font[text[l]-32][0]); // -32 is to get to correct ascii
  sendPixelRow(font[text[l]-32][1]); // -32 is to get to correct ascii
  sendPixelRow(font[text[l]-32][2]); // -32 is to get to correct ascii
  sendPixelRow(font[text[l]-32][3]); // -32 is to get to correct ascii
  sendPixelRow(font[text[l]-32][4]); // -32 is to get to correct ascii
  sendPixelRow(font[text[l]-32][5]); // -32 is to get to correct ascii
}
```

```
  sei();
  delay(4000);              // Wait to display results

  cli();                    // No time for interruptions!
  for (int l=0; l<maxled; l++){    // 30 is number of characters to clear
    sendPixelRow(0x00);     // clear the screen
  }
  sei();

// FIREWORKS
  cli();                    // No time for interruptions!
  red=textb; green=textb; blue=textb;
  for (int cycle=0; cycle<100; cycle++){
    int rabit=random(7);
    byte ls=0x01;
    byte sbit=ls << rabit;  // Set up random bit
    int rano=random(maxled);    // Set Random location
    for (int l=0; l<maxled; l++){    // 180 is number of bytes to send
      if (l==rano) sendPixelRow(sbit); else sendPixelRow(0x00);
    }
    delay(20);
  }
  sei();
  return;
}
```

When you examine the code you will first see that I am sending the same data out the data port several times. Each time it takes about 100 ns, so sending the same thing four times takes about 400ns.

I reduced the font file so there are no lower case letters in the listing. Again I am trying to save time and to keep the program listing short. You can add back lower case letters but that might make the lookup of characters take too long.

When selecting the color for the letters I used a Switch/Case instead of using several if's. All if statements must be evaluated every time but a switch just goes to the correct value taking much less time. So the Switch is much faster than using several if statements.

Then when sending the five rows of each of the characters out I did not use a loop like: "for (int r=0; r<6; r++){ " because using a loop doubles the amount of time that is used. Also coding the row bytes as five statements is not that much longer.

Scrolling Text Sign Program

If you want to scroll the text you need to give up the ability to change the color for each character. Unless you want to have three arrays, one array for each color. What I did was to load all of the selected characters into an array called texts. Then the array is padded with 30 blank characters. Next the array is sent to the LED's but it is offset by a variable.

Here is the code for a scrolling sign:
```
// Version for sending data to 8 Parallel WS2812 strings
// by: BOB Davis
// Scrolling Text Version

// PORTD is Digital Pins 0-7 on the Uno change for other boards.
#define PIXEL_PORT  PORTD  // Port the pixels are connected to
#define PIXEL_DDR   DDRD   // Port the pixels are connected to
int textb=128;    // brightness of text
int maxled=180;   // Number of LED's
byte texts[360]; // Characters times Columns (30x6=180)*2
char text[]="   ARDUINO LED STRIP PROJECTS          ";
// Set default color for letters
int red=1; int green=textb; int blue=1;

// Send the next set of 8 WS2812B encoded bits to the 8 pins.
// The delay timing is for an Arduino UNO.
 void sendBitX8( uint8_t bits ) {
  PORTD= 0xFF;  // turn on
  PORTD= 0xFF;  // delay
  PORTD= 0xFF;  // delay
  PORTD= 0xFF;  // delay
```

```
  PORTD= 0xFF;  // delay (add more for faster processors)
  PORTD= bits;  // send data
  PORTD= bits;  // delay
  PORTD= bits;  // delay
  PORTD= bits;  // delay
  PORTD= bits;  // delay
  PORTD= 0x00;  // Turn off;
}

void sendPixelRow( uint8_t row ) {
  // Send the bit 8 times down every row, each pixel is 8 bits/color
  int mask = 0x01;  // shifting mask to determine bit status
    for (int bit=8; bit>0; bit--){
      if (green & (mask<<bit))sendBitX8( row ); else sendBitX8( 0x00 ); }
    for (int bit=8; bit>0; bit--){
      if (red & (mask<<bit))sendBitX8( row ); else sendBitX8( 0x00 ); }
    for (int bit=8; bit>0; bit--){
      if (blue & (mask<<bit))sendBitX8( row ); else sendBitX8( 0x00 ); }
  }

// Font from http://sunge.awardspace.com/glcd-sd/node4.html
byte font[][6] = {
0x00,0x00,0x00,0x00,0x00,0x00, // ascii 32
0x00,0x00,0xfa,0x00,0x00,0x00, // !
0x00,0xe0,0x00,0xe0,0x00,0x00, // "
0x28,0xfe,0x28,0xfe,0x28,0x00, // #
0x24,0x54,0xfe,0x54,0x48,0x00, // $
0xc4,0xc8,0x10,0x26,0x46,0x00, // %
0x6c,0x92,0xaa,0x44,0x0a,0x00, // &
0x00,0xa0,0xc0,0x00,0x00,0x00, // '
0x00,0x38,0x44,0x82,0x00,0x00, // (
0x00,0x82,0x44,0x38,0x00,0x00, // )
0x10,0x54,0x38,0x54,0x10,0x00, // *
0x10,0x10,0x7c,0x10,0x10,0x00, // +
0x00,0x0a,0x0c,0x00,0x00,0x00, // ,
0x10,0x10,0x10,0x10,0x10,0x00, // -
```

```
0x00,0x06,0x06,0x00,0x00,0x00, // .
0x04,0x08,0x10,0x20,0x40,0x00, // /
0x7c,0x8a,0x92,0xa2,0x7c,0x00, // 0
0x00,0x42,0xfe,0x02,0x00,0x00, // 1
0x42,0x86,0x8a,0x92,0x62,0x00, // 2
0x84,0x82,0xa2,0xd2,0x8c,0x00, // 3
0x18,0x28,0x48,0xfe,0x08,0x00, // 4
0xe4,0xa2,0xa2,0xa2,0x9c,0x00, // 5
0x3c,0x52,0x92,0x92,0x0c,0x00, // 6
0x80,0x8e,0x90,0xa0,0xc0,0x00, // 7
0x6c,0x92,0x92,0x92,0x6c,0x00, // 8
0x60,0x92,0x92,0x94,0x78,0x00, // 9
0x00,0x6c,0x6c,0x00,0x00,0x00, // :
0x00,0x6a,0x6c,0x00,0x00,0x00, // ;
0x00,0x10,0x28,0x44,0x82,0x00, // <
0x28,0x28,0x28,0x28,0x28,0x00, // =
0x82,0x44,0x28,0x10,0x00,0x00, // >
0x40,0x80,0x8a,0x90,0x60,0x00, // ?
0x4c,0x92,0x9e,0x82,0x7c,0x00, // @
0x7e,0x90,0x90,0x90,0x7e,0x00, // A
0xfe,0x92,0x92,0x92,0x6c,0x00, // B
0x7c,0x82,0x82,0x82,0x44,0x00, // C
0xfe,0x82,0x82,0x44,0x38,0x00, // D
0xfe,0x92,0x92,0x92,0x82,0x00, // E
0xfe,0x90,0x90,0x80,0x80,0x00, // F
0x7c,0x82,0x82,0x8a,0x4c,0x00, // G
0xfe,0x10,0x10,0x10,0xfe,0x00, // H
0x00,0x82,0xfe,0x82,0x00,0x00, // I
0x04,0x02,0x82,0xfc,0x80,0x00, // J
0xfe,0x10,0x28,0x44,0x82,0x00, // K
0xfe,0x02,0x02,0x02,0x02,0x00, // L
0xfe,0x40,0x20,0x40,0xfe,0x00, // M
0xfe,0x20,0x10,0x08,0xfe,0x00, // N
0x7c,0x82,0x82,0x82,0x7c,0x00, // O
0xfe,0x90,0x90,0x90,0x60,0x00, // P
0x7c,0x82,0x8a,0x84,0x7a,0x00, // Q
```

```
0xfe,0x90,0x98,0x94,0x62,0x00, // R
0x62,0x92,0x92,0x92,0x8c,0x00, // S
0x80,0x80,0xfe,0x80,0x80,0x00, // T
0xfc,0x02,0x02,0x02,0xfc,0x00, // U
0xf8,0x04,0x02,0x04,0xf8,0x00, // V
0xfe,0x04,0x18,0x04,0xfe,0x00, // W
0xc6,0x28,0x10,0x28,0xc6,0x00, // X
0xc0,0x20,0x1e,0x20,0xc0,0x00, // Y
0x86,0x8a,0x92,0xa2,0xc2,0x00, // Z
};

void setup() {
  PIXEL_DDR = 0xff;   // Set all row pins to output
}
void loop() {
  red=1; green=textb; blue=1;
  // Load Array
  for (int l=0; l<30; l++){    // 30 is number of characters to send
    for (int r=0; r<6; r++){    // 6 rows per character
      texts[(l*6)+r]=(font[text[l]-32][r]); // -32 is to get to correct ascii
    }
  }
  for (int l=0; l<30; l++){    // 30 is number of characters to send
    for (int r=0; r<6; r++){    // 6 rows per character
      texts[(l*6+maxled)+r]=0x00;    // Blanks otherwise see memory contents
    }
  }

  // Display Array
  cli();                // No time for interruptions!
  for (int s=0; s<maxled; s++){   // s in number to skip
    for (int l=0; l<maxled; l++){    // number of LEDs to send
      sendPixelRow(texts[l+s]);  // send array to display
    }
    delay(5000);
```

```
  }
  sei();

  cli();                  // No time for interruptions!
  for (int l=0; l<maxled; l++){   // number of LEDs to clear
    sendPixelRow(0x00);   // clear the screen
  }
  sei();

  // FIREWORKS
  red=textb; green=textb; blue=textb;
  cli();                  // No time for interruptions!
  for (int cycle=0; cycle<100; cycle++){
    int rabit=random(7);
    byte ls=0x01;
    byte sbit=ls << rabit;    // Set up random bit
    int rano=random(maxled);  // Set Random location
    for (int l=0; l<maxled; l++){  // 180 is number of bytes to send
      if (l==rano) sendPixelRow(sbit); else sendPixelRow(0x00);
    }
    delay(20);
  }
  sei();
  return;
}
```

Here is an alternate, more intense fireworks routine. It stores a random number for each position across the sign. So instead of one flash per sign there is a flash per vertical position on the sign. The settings shown are for two rows of 90 LED's each. You can adjust them for your sign. If you change the seed number of 0x00001 to 0x00003 it looks a little like a fire.

```
void loop() {
// More intense fireworks
  int rnum[180];
  int solidt[180];
```

```
    int solidb[180];
    cli();              // No time for interruptions!
    for (int n=0; n<maxled/2; n++){
      rnum[n]=random(16); // random 0 to 15
      // processing must be done here it takes too long in display loop
      int ls=0x00001;
      solidb[n]=ls << rnum[n];
      solidt[n]=solidb[n] >> 8;
      }
    red=textb; green=0; blue=0;
    for (int l=0; l<maxled/2; l++){  // number of bytes to send
      sendPixelRow(solidt[l]);
    }
    for (int l=0; l<maxled/2; l++){  // number of bytes to send
      sendPixelRow(solidb[l]);
    }
    delay(5000);
    sei();
      return;
}
```

With a few more changes you can make the random numbers fall to the bottom. It looks a little bit like falling rain.

```
void loop() {
// Falling Random Numbers
  int rnum[180];
  int solidt[180];
  int solidb[180];
  int replnum;
  for (int n=0; n<maxled/2; n++){
    rnum[n]=random(16); // random 0 to 15
  }
  for (int c=0; c<1000; c++){
    for (int n=0; n<maxled/2; n++){
      rnum[n]=rnum[n]-1; // move numbers down
```

```
  }
  for (int nn=0; nn<maxled/10; nn++){
    replnum=random(90);
    rnum[replnum]=random(16);  // randomly replace numbers
  }
  for (int n=0; n<maxled/2; n++){
    // processing must be done here it takes too long in display loop
    int ls=0x00001;
    solidb[n]=ls << rnum[n];
    solidt[n]=solidb[n] >> 8;
  }
  red=0;  green=textb;  blue=0;
  cli();                  // No time for interruptions!
  for (int l=0; l<maxled/2; l++){   // number of bytes to send
    sendPixelRow(solidt[l]);
  }
  for (int l=0; l<maxled/2; l++){   // number of bytes to send
    sendPixelRow(solidb[l]);
  }
 delay(5000);
 }
 sei();
}
```

Chapter 4

Adding a MSEQ7 for More Effects

You can make an audio spectrum analyzer using an Arduino UNO a MSGEQ7 and some WS2812 LED Strips. The audio spectrum display can be as big as you want because the software supports strips of 400 or more LED's.

The MSGEQ7 is a little eight pin chip that has seven band pass amplifiers, seven peak detectors and an output multiplexer. It is amazing that someone thought to stuff all of that technology into one IC. The next picture shows its pin assignments.

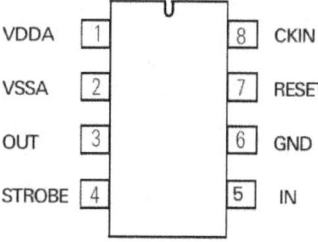

The biggest complaint about the MSGEQ7 is that it only has seven bands. Most equalizers have 10, 15, or more bands. In the MSGEQ7 each band is the previous band divided by 2.5. However it is missing the 25 hertz band at the low end. We can always add that band with a second IC. Here is a picture of the seven frequency bands of the MSGEQ7 from the specification sheet.

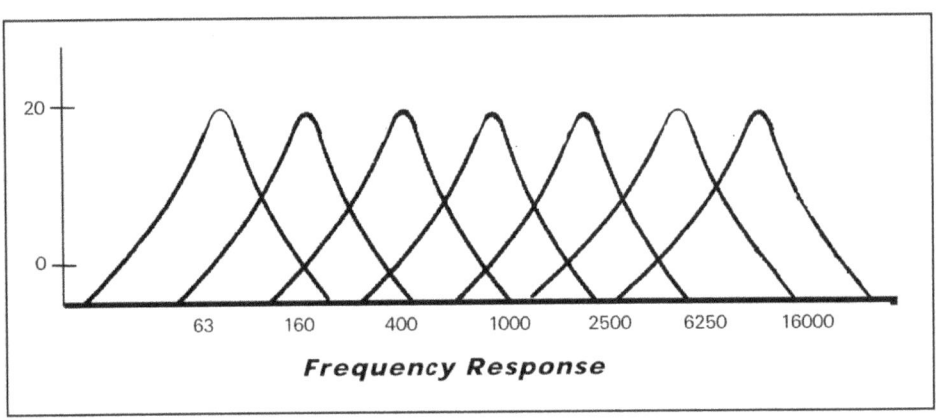

Frequency Response

After lots of failures I was finally been able to get the combination of a MSGEQ7 and WS2812 LED strips to work together. The trick was to not use any machine language code to send data to the WS2812 LED strips. Instead, to access the WS2812 I just used parallel output commands and then repeated the commands to obtain a time delay. Sending the same commands over and over just results in a time delay of about 100ns each.

I also modified my MSGEQ7 shield. I changed the input resistors (R3 and R4) from 22K to 100 ohms. The lower resistance works better with headphone level inputs and helps reduce the amount of noise picked up due to the lower impedance. This next picture shows my modified MSGEQ7 shield. Newer versions of this shield have a prototype area.

Here is the MSGEQ7 Shield schematic diagram. This is the Chinese shield, it has a different pin arrangement from the American version.

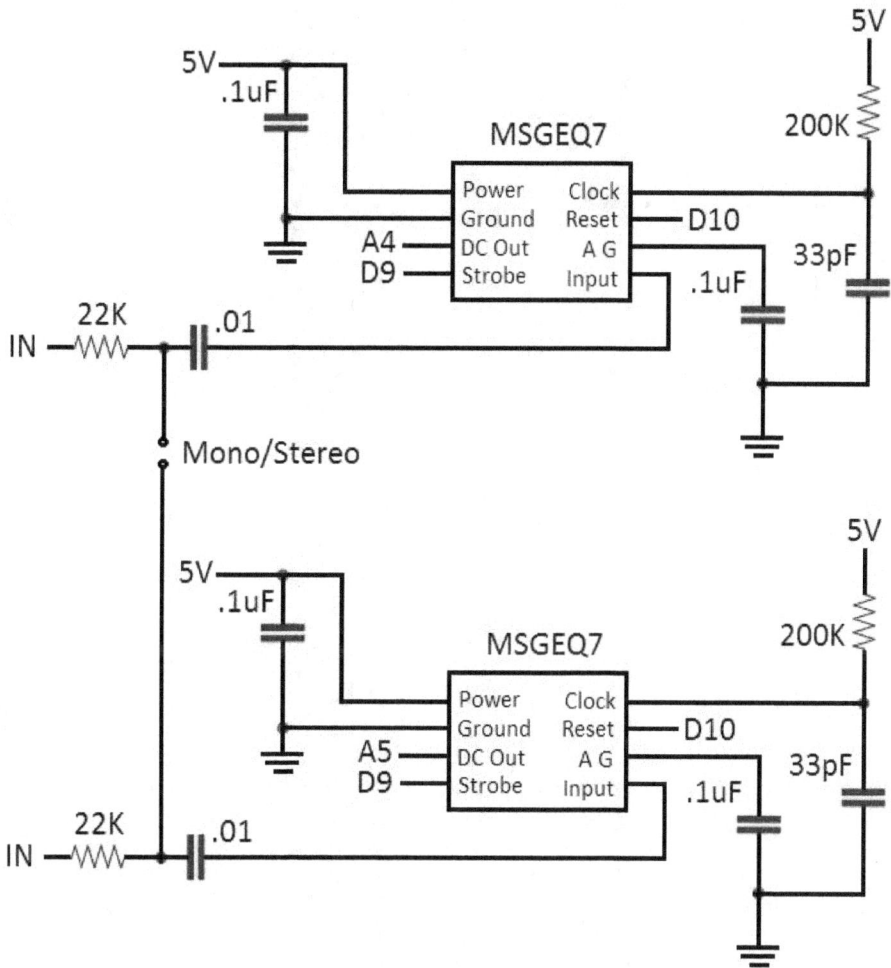

Audio Spectrum Analyzer

My software gives you 90 steps or LED's to the top of the sign board. Of course you can go much bigger if you want to but the ceiling height can be a limiting factor.

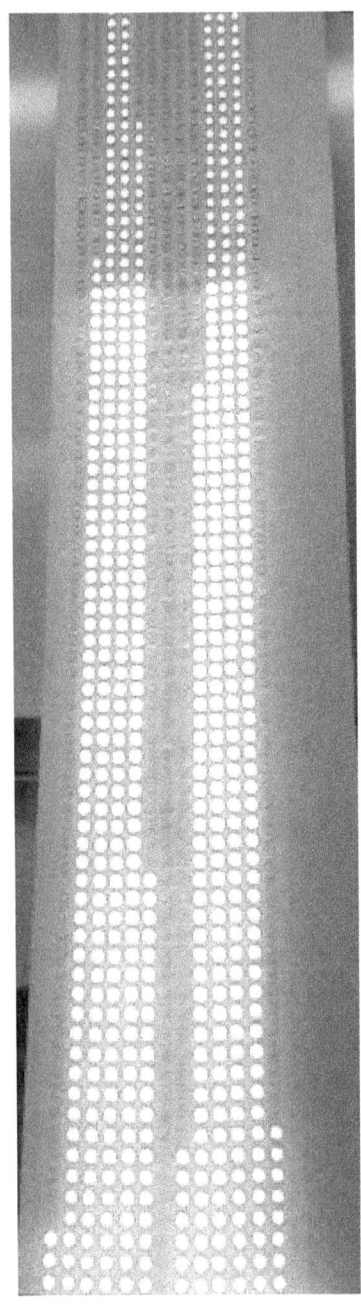

This is the code for the right and left channel version. The second channel starts after 90 LED's. You might need to adjust it to the amount of LED's you have in each sign board.

What looks a little bit confusing is the code to turn on the binary bit to match with the column of the display. Instead of adding 2, 4, 8 etc. you could or 0x02, 0x04, 0x08, etc.

```
// Version for sending data to 8 Parallel WS2812 strings
// By Bob Davis
// Removed Assembler and simplified the code
// Changed to VU meter with MSGEQ7

// PORTD is Digital Pins 0-7 on the Uno change for other boards.
#define PIXEL_PORT  PORTD  // Port connected to
#define PIXEL_DDR   DDRD   // Port connected to
int textb=128;  // brightness of text
int maxled=180; // Number of LED's

// MSGEQ7 pins
#define PIN_STROBE 9
#define PIN_RESET 10
#define PIN_LEFT 4 //analog input
#define PIN_RIGHT 5 //analog input

//band arrays
int left[7];
int right[7];
int col=0;

void readMSGEQ7() { //reset the chip
  digitalWrite(PIN_RESET, HIGH);
  digitalWrite(PIN_RESET, LOW);
  for(int band=0; band < 7; band++) { //loop thru all 7 bands
    digitalWrite(PIN_STROBE,LOW);     // go to the next band
    delayMicroseconds(30);            // gather data
    left[band] = analogRead(PIN_LEFT); // store left band reading
    right[band] = analogRead(PIN_RIGHT); // store right band reading
    digitalWrite(PIN_STROBE,HIGH);    // reset the strobe pin
  }
}
```

}

```
// Actually send the next set of 8 WS2812B encoded bits to the 8 pins.
// The delay timing is for an Arduino UNO.
void sendBitX8( uint8_t bits ) {
  PORTD= 0xFF;  // turn on
  PORTD= 0xFF;  // delay
  PORTD= 0xFF;  // delay
  PORTD= 0xFF;  // delay
  PORTD= 0xFF;  // delay (Remove for faster processors)
  PORTD= bits;  // send data
  PORTD= bits;  // delay
  PORTD= bits;  // delay
  PORTD= bits;  // delay
  PORTD= bits;  // delay
  PORTD= 0x00;  // Turn off;
}

// Set default color for letters
int red=1; int green=1; int blue=1;

void sendPixelRow( uint8_t row ) {
  // Send the bit 8 times down every row, each pixel is 8 bits/color
  int mask = 0x01;  // shifting mask to determine bit status
    for (int bit=8; bit>0; bit--){
      if (green & (mask<<bit))sendBitX8( row ); else sendBitX8( 0x00 ); }
    for (int bit=8; bit>0; bit--){
      if (red & (mask<<bit))sendBitX8( row ); else sendBitX8( 0x00 ); }
    for (int bit=8; bit>0; bit--){
      if (blue & (mask<<bit))sendBitX8( row ); else sendBitX8( 0x00 ); }
  }

void setup() {
  PIXEL_DDR = 0xff;   // Set all row pins to output
  pinMode(PIN_RESET, OUTPUT); // reset
  pinMode(PIN_STROBE, OUTPUT); // strobe
}

void loop() {
    readMSGEQ7();            // collect samples
    cli();                   // No time for interruptions!
```

```
  for (int b=0; b<maxled/2; b++){    //90 is number of bytes to send
   if (b < maxled/4) {red=0; green=textb; blue=0;}
   if (b > maxled/4+10) {red=textb; green=textb;}
   if (b > maxled/4+20) {red=textb; green=0;}
   col=0;
   if (left[0]-128 >= b) col=col+2;   // Send bytes as VU meter data
   if (left[1]-128 >= b) col=col+4;   // Send bytes as VU meter data
   if (left[2]-128 >= b) col=col+8;   // Send bytes as VU meter data
   if (left[3]-128 >= b) col=col+16;  // Send bytes as VU meter data
   if (left[4]-128 >= b) col=col+32;  // Send bytes as VU meter data
   if (left[5]-128 >= b) col=col+64;  // Send bytes as VU meter data
   if (left[6]-128 >= b) col=col+128; // Send bytes as VU meter data
   sendPixelRow(col);
   }
  for (int b=0; b<maxled/2; b++){    //90 is number of bytes to send
   if (b < maxled/4) {red=0; green=textb; blue=0;}
   if (b > maxled/4+10) {red=textb; green=textb;}
   if (b > maxled/4+20) {red=textb; green=0;}
   col=0;
   if (right[0]-128 >= b) col=col+2;   // Send bytes as VU meter data
   if (right[1]-128 >= b) col=col+4;   // Send bytes as VU meter data
   if (right[2]-128 >= b) col=col+8;   // Send bytes as VU meter data
   if (right[3]-128 >= b) col=col+16;  // Send bytes as VU meter data
   if (right[4]-128 >= b) col=col+32;  // Send bytes as VU meter data
   if (right[5]-128 >= b) col=col+64;  // Send bytes as VU meter data
   if (right[6]-128 >= b) col=col+128; // Send bytes as VU meter data
   sendPixelRow(col);
   }
  sei();             // interrupts back on
  delay (50);
}
```

42 Frequency Spectrum Analyzer

If you want more than seven frequencies displayed per channel it is possible to offset the clock of the MSGEQ7's. To do that you replace the 200K resistor and 33 pf capacitor with different values. The resistor and capacitor are located on the right side of the previous MSGEQ7 schematic diagram. Increasing their values decreases the frequency. The MSGEQ7 specification sheet only states a 10% tolerance is acceptable but the reality is much wider variations will usually work.

A 60 percent shift is recommended to reduce the overlap of frequencies. An example of that would be to use 220K and 47pf for 60 percent change in values. The lowest frequency displayed is now about 40 hertz instead of 63 hertz. Changing the values 60 percent changes the frequency about 30 percent. If you were to decrease the resistor and capacitor 60 percent the highest frequency would then be above 20KHz! That is too high to be useable. Remember that the MSGEQ7 uses a multiple of 2.5 between each frequency. That is a 250 percent change.

You can use several offset MSGEQ7's to get as many frequencies as you like, but three per channel is a practical limit. That is because the Arduino Uno had six analog inputs for three MSGEQ7's per channel for a total of six of them. Also you would have to modify some shields to change the capacitor and resistor values and to change the analog input pins or you can make your own MSGEQ7 shield for the Arduino using the previous schematic.

To display more than 14 frequencies it is best to turn the LED display back right side up or horizontal. That limits you to 8 steps per frequency but you can now display a lot more frequencies.

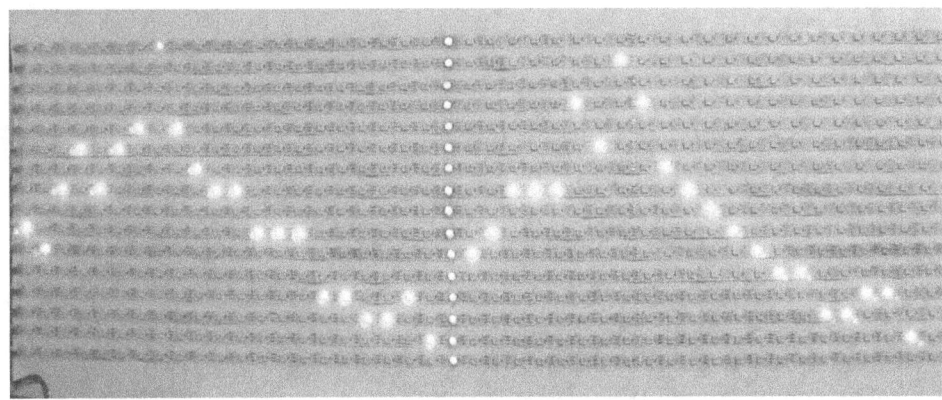

This next program supports 3 MSGEQ7's per channel for a total of 21 frequencies per channel or 42 all together.

```
// Version for sending data to 8 Parallel WS2812 strings
// By BOB Davis
// Changed to VU meter with MSGEQ7 Supports 6 MSGEQ7's
// Displays dots on 16 lines of LED's.

// PORTD is Digital Pins 0-7 on the Uno change for other boards.
#define PIXEL_PORT  PORTD // Port they are connected to
#define PIXEL_DDR   DDRD  // Port they are connected to
int textb=128;  // Brightness of text
int maxled=180; // Number of LED's

// MSGEQ7 pins
#define PIN_STROBE 9
#define PIN_RESET 10
#define PIN_0 0 //analog input
#define PIN_1 1 //analog input
#define PIN_2 2 //analog input
#define PIN_3 3 //analog input
#define PIN_4 4 //analog input
#define PIN_5 5 //analog input

//band arrays
int left0[7];
int right0[7];
int left1[7];
int right1[7];
int left2[7];
```

```
int right2[7];
int col=0;

void readMSGEQ7() { //reset the chip
  digitalWrite(PIN_RESET, HIGH);
  digitalWrite(PIN_RESET, LOW);
  for(int band=0; band < 7; band++) { //loop thru all 7 bands
    digitalWrite(PIN_STROBE,LOW);    // go to the next band
    delayMicroseconds(30);           // gather data
    left0[band] = analogRead(PIN_0)-32;  // store band reading
    right0[band] = analogRead(PIN_1)-32; // store band reading
    left1[band] = analogRead(PIN_2)-32;  // store band reading
    right1[band] = analogRead(PIN_3)-32; // store band reading
    left2[band] = analogRead(PIN_4)-32;  // store band reading
    right2[band] = analogRead(PIN_5)-32; // store band reading
    digitalWrite(PIN_STROBE,HIGH);   // reset the strobe pin
  }
}

// Actually send the next set of 8 WS2812B encoded bits to the 8 pins.
// The delay timing is for an Arduino UNO.
void sendBitX8( uint8_t bits ) {
  PORTD= 0xFF;  // turn on
  PORTD= 0xFF;  // delay
  PORTD= 0xFF;  // delay
  PORTD= 0xFF;  // delay
  PORTD= 0xFF;  // delay (Increase for faster processors)
  PORTD= bits;  // send data
  PORTD= bits;  // delay
  PORTD= bits;  // delay
  PORTD= bits;  // delay
  PORTD= bits;  // delay
  PORTD= 0x00;  // Turn off;
}

// Set default color for letters
int red=1; int green=textb; int blue=1;

void sendPixelRow( uint8_t row ) {
  // Send the bit 8 times down every row, each pixel is 8 bits/color
  int mask = 0x01;  // shifting mask to determine bit status
```

```
    for (int bit=8; bit>0; bit--){
      if (green & (mask<<bit))sendBitX8( row ); else sendBitX8( 0x00 ); }
    for (int bit=8; bit>0; bit--){
      if (red & (mask<<bit))sendBitX8( row ); else sendBitX8( 0x00 ); }
    for (int bit=8; bit>0; bit--){
      if (blue & (mask<<bit))sendBitX8( row ); else sendBitX8( 0x00 ); }
}

void setup() {
  PIXEL_DDR = 0xff;   // Set all row pins to output
  pinMode(PIN_RESET, OUTPUT); // reset
  pinMode(PIN_STROBE, OUTPUT); // strobe
}

void loop() {
   readMSGEQ7();        // collect samples
   cli();              // No time for interruptions!
   // Top half of display
   for (int b=0; b<7; b++){  //7 is number of freq per MSGEQ7
     col=0x01;            // Set initial value
     col=col<<left0[b]/64;   // Shift active bit up to match level
     col=col>>8;
     if (col&0x80) {red=textb; green=0;}
     else {red=0; green=textb;}
     sendPixelRow(col);
     col=0x01;            // Set initial value
     col=col<<left1[b]/64;   // Shift active bit up to match level
     col=col>>8;
     if (col&0x80) {red=textb; green=0;}
     else {red=0; green=textb;}
     sendPixelRow(col);
     col=0x01;            // Set initial value
     col=col<<left2[b]/64;   // Shift active bit up to match level
     col=col>>8;
     if (col&0x80) {red=textb; green=0;}
     else {red=0; green=textb;}
     sendPixelRow(col);
   }
   red=0; green=0; blue=textb;
   sendPixelRow(0xFF);     // Left to Right separator marker.
   red=0; green=textb; blue=0;
```

```
for (int b=0; b<7; b++){  //7 is number of freq per MSGEQ7
  col=0x01;              // Set initial value
  col=col<<right0[b]/64;  // Shift active bit up to match level
  col=col>>8;
  if (col&0x80) {red=textb; green=0;}
  else {red=0; green=textb;}
  sendPixelRow(col);
  col=0x01;              // Set initial value
  col=col<<right1[b]/64;  // Shift active bit up to match level
  col=col>>8;

  if (col&0x80) {red=textb; green=0;}
  else {red=0; green=textb;}
  sendPixelRow(col);
  col=0x01;              // Set initial value
  col=col<<right2[b]/64;  // Shift active bit up to match level
  col=col>>8;

  if (col&0x80) {red=textb; green=0;}
  else {red=0; green=textb;}
  sendPixelRow(col);
}
int space=47;  // number of blanks to roll over to next line
col=0x00;
for (int s=0; s<space; s++) {sendPixelRow(col);}

//  Bottom half of display
red=0; green=textb;
for (int b=0; b<7; b++){  //7 is number of freq per MSGEQ7
  col=0x01;              // Set initial value
  col=col<<left0[b]/64;   // Shift active bit up to match level
  sendPixelRow(col);
  col=0x01;              // Set initial value
  col=col<<left1[b]/64;   // Shift active bit up to match level
  sendPixelRow(col);
  col=0x01;              // Set initial value
  col=col<<left2[b]/64;   // Shift active bit up to match level
  sendPixelRow(col);
}
red=0; green=0; blue=textb;
sendPixelRow(0xFF);       // Left to Right separator marker.
```

```
red=0; green=textb; blue=0;
for (int b=0; b<7; b++){   //7 is number of freq per MSGEQ7
  col=0x01;              // Set initial value
  col=col<<right0[b]/64;   // Shift active bit up to match level
  sendPixelRow(col);
  col=0x01;              // Set initial value
  col=col<<right1[b]/64;   // Shift active bit up to match level
  sendPixelRow(col);
  col=0x01;              // Set initial value
  col=col<<right2[b]/64;   // Shift active bit up to match level
  sendPixelRow(col);
}
  sei();              // interrupts back on
  delay (50);
}
```

If you prefer bar-graphs instead of dots you will need to get a little tricky. There is no left shift that automatically brings in 1's for the Arduino that I know of. So we start off with 0x00FF and shifted that left for the value of the audio level. Then we shift it right to dump off any extra 1's that were not previously shifted left.

This next program is only the loop portion of the code as the rest of the code stays essentially the same. It also is for eight lines of LED's only as making 16 lines just takes twice as much code and I would assume even some more shifting.

```
void loop() {
  readMSGEQ7();          // collect samples
  cli();              // No time for interruptions!
  for (int b=0; b<7; b++){   //7 is number of freq per MSGEQ7
    col=0x00FF;            // Set initial value
    col=col<<left0[b]/64;   // Shift active bit up to match level
    col=col>>8;          // Extra shift dumps the noise
    if (col&0x80) {red=textb; green=0;}
    else {red=0; green=textb;}
    sendPixelRow(col);
    col=0x00FF;            // Set initial value
    col=col<<left1[b]/64;   // Shift active bit up to match level
    col=col>>8;
    if (col&0x80) {red=textb; green=0;}
```

```
      else {red=0; green=textb;}
      sendPixelRow(col);
      col=0x00FF;             // Set initial value
      col=col<<left2[b]/64;   // Shift active bit up to match level
      col=col>>8;
      if (col&0x80) {red=textb; green=0;}
      else {red=0; green=textb;}
      sendPixelRow(col);
    }
    red=0; green=0; blue=textb;
    sendPixelRow(0xFF);       // Left to Right separator marker.
    red=0; green=textb; blue=0;
    for (int b=0; b<7; b++){  //7 is number of freq per MSGEQ7
      col=0x00FF;             // Set initial value
      col=col<<right0[b]/64;  // Shift active bit up to match level
      col=col>>8;
      if (col&0x80) {red=textb; green=0;}
      else {red=0; green=textb;}
      sendPixelRow(col);
      col=0x00FF;             // Set initial value
      col=col<<right1[b]/64;  // Shift active bit up to match level
      col=col>>8;
      if (col&0x80) {red=textb; green=0;}
      else {red=0; green=textb;}
      sendPixelRow(col);
      col=0x00FF;             // Set initial value
      col=col<<right2[b]/64;  // Shift active bit up to match level
      col=col>>8;
      if (col&0x80) {red=textb; green=0;}
      else {red=0; green=textb;}
      sendPixelRow(col);
    }
    sei();                    // interrupts back on
    delay (50);
```

Oscilloscope Like Display

To make the sign into an oscilloscope like display all you need to do is add a couple of definitions like this next code sample. We are going to collect several audio samples then store them to display them later.

```
#define APIN 0  //analog input
int right[200];  // number of samples
int col=0;
```

Then change the "loop" to look like this code listing. We collect samples equal to the width of the sign. A slight delay is needed for normal audio signals. Then we display the top half of the signal and then the bottom half is displayed.

```
void loop() {
   for (int s=0; s<maxled/2; s++){
   right[s]=analogRead(APIN);         // collect samples
   delay(1);
   }
   cli();               // No time for interruptions!
   for (int b=0; b<(maxled/2); b++){
    col=0x01;               // Set initial value
    col=col<<right[b]/32;   // Shift active bit up to match level
    col=col>>8;
    if (col&0x80) {red=textb; green=0;}
    else {red=0; green=textb;}
    sendPixelRow(col);
   }
   red=0; green=textb; blue=0;
   //  Bottom half
   for (int b=0; b<(maxled/2); b++){
    col=0x01;               // Set initial value
    col=col<<right[b]/32;   // Shift active bit up to match level
    if (col&0x01) {red=textb; green=0;}
    else {red=0; green=textb;}
    sendPixelRow(col);
   }
   sei();              // interrupts back on
   delay (50);
}
```

The audio should go through a coupling capacitor at a very minimum. However a front end with a centering or position capability works the best. Also adding a zener diode will protect the analog input from any spikes in power. This next schematic diagram shows how to make a simple front end with a position control.

```
              100
    Input─┤├─/\/\/\──────────── To Analog
      .47 uF         │    ⌇5V   Input Pin.
                   100K
                     │
              5V ─/\/\/─── Ground
                  10 K
                 Position
```

This picture shows what the oscilloscope display looks like when it is displaying a 100 hertz audio sine wave. If the signal is too strong the edge LED's will turn red.

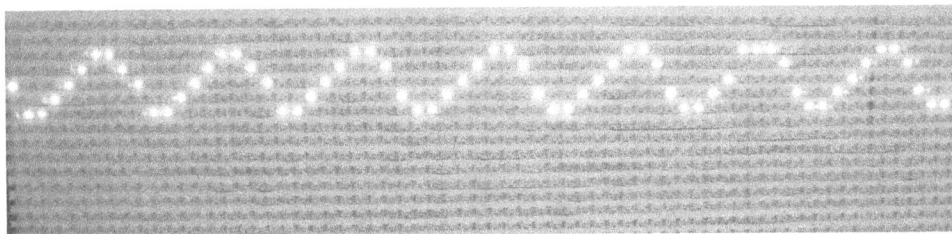

Chapter 5

Adding Serial Communication

Adding serial communication to the LED sign can be tricky. However doing that will open the door to USB, Bluetooth and other types of wireless communication with the sign. The problem is that we have been using D0 and D1 for the sign. They are by default also used for serial communication. That is why the bottom two rows of LED's do strange things when we are updating the software.

What we will have to do is shift some of the data bits to port B. In fact we will leave the higher four bits on port D but the lower four bits will now be on port B. You could also use this second port to send as many as 14 bits to a sign controlling 14 rows of LED strips. That would allow even bigger fonts, etc. But for now we are just doing it so we can use data pins D0 and D1 for serial communications.

If the connector that plugs into the Arduino Uno is an eight pin connector it would have to be broken in half so it is two pieces of four pins each.

The schematic diagram would have to be changed to this next image to use both ports B and D. We are talking about port B, but the Arduino connections are labeled as Data 0 to Data 13. To translate from port to data pin add 8 to the port bit. So then B0 is D8, B1 is D9, B2 is D10 and B3 is D11.

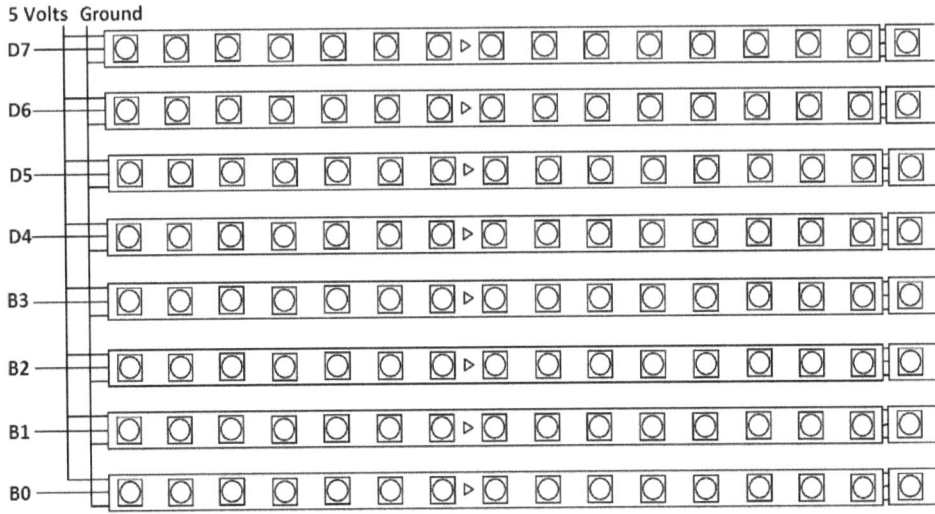

To communicate with the sign over the USB cable, in the Arduino IDE select "Tools" and "Serial Monitor". A box will come up where on the top line you can type in text to send and the larger area at the bottom shows what came back from the Arduino.

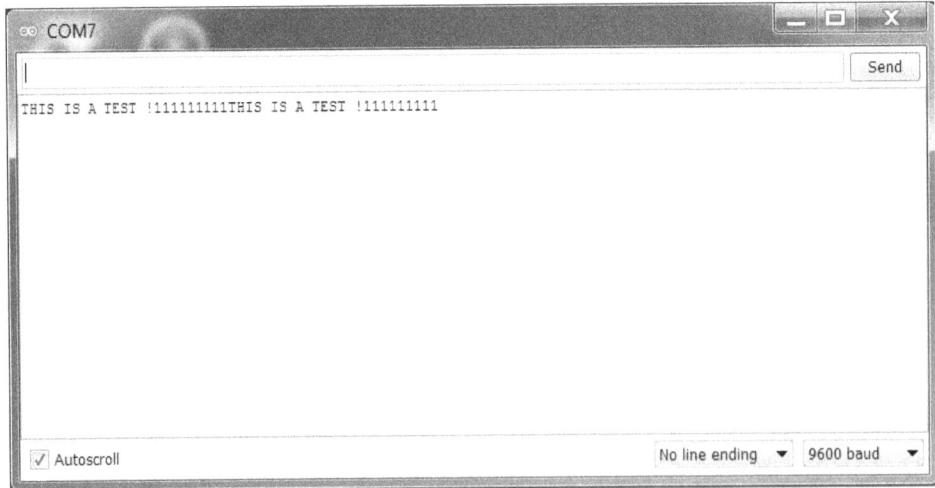

Once you can get USB communication working you can then advance to wireless communications devices. One wireless method uses a 2.4 GHz radio frequency communication that is called "Bluetooth". Probably the best wireless device to use for controlling things is "Bluetooth" because you can also control it with your telephone or anything else that can send data to a Bluetooth device. Bluetooth also emulates a serial port, so data can be sent and received just like it was a direct serial or USB connection.

When the Bluetooth receiver arrived in the mail it did not came with any real instructions. All that I had to work with was these specifications:

Default serial port setting : 9600 1
Pairing code: 1234
Running in slave role : Pair with BT dongle and master module
Coupled Mode : Two modules will establish communication automatically when they are powered up.
PC hosted mode : Pair the module with Bluetooth dongle directly as a virtual serial device.
Bluetooth protocol : Bluetooth Specification v2.0+EDR
Frequency : 2.4GHz ISM band
Modulation : GFSK(Gaussian Frequency Shift Keying)
Emission power : <=4dBm, Class 2
Sensitivity : <=-84dBm at 0.1% BER
Speed : Asynchronous: 2.1Mbps(Max) / 160 kbps,
Synchronous : 1Mbps/1Mbps
Security : Authentication and encryption
Profiles : Bluetooth serial port
CSR chip : Bluetooth v2.0
Wave band : 2.4GHz-2.8GHz, ISM Band
Protocol : Bluetooth V2.0
Power Class : (+6dbm)
Reception sensitivity: -85dBm
Voltage : 3.3 (2.7V-4.2V)
Current : Paring - 35mA, Connected - 8mA
Temperature : -40~ +105 Degrees Celsius
User defined Baud rate : 4800, 9600, 19200, 38400, 57600, 115200, 230400,460800,921600 ,1382400.
Dimension : 26.9mm*13mm*2.2mm

Setting up a Bluetooth wireless remote control is not an easy thing to do the first time. It took me several hours to get it working properly. First, start by plugging in the USB Bluetooth adapter to your PC's USB jack and then installing the needed drivers. Usually it will automatically install the drivers over the Internet if you select that option and have a working Internet connection.

Next you can power up the Bluetooth receiver module that will connect to the Arduino. Only connect the power and ground pins to the receiver for now. If you have the complete adapter with a voltage regulator built in it will work on either 3.3 or 5 volts, otherwise it must have 3.3 volts to operate. The receiver should start blinking a red LED to let you know that it has power.

On the PC, go to "Control panel" and select the "Bluetooth devices" icon. In the menu that comes up, select "add" and then check the box that says "It is powered up and ready to be found" and then select "next". Then you should see as the computer searching for Bluetooth devices.

Most of the Arduino interface Bluetooth devices will come up as "HC-06". Select it, and at the prompt, enter the default pass code of "1234". The "Add Bluetooth Device" screen then look something like the following picture.

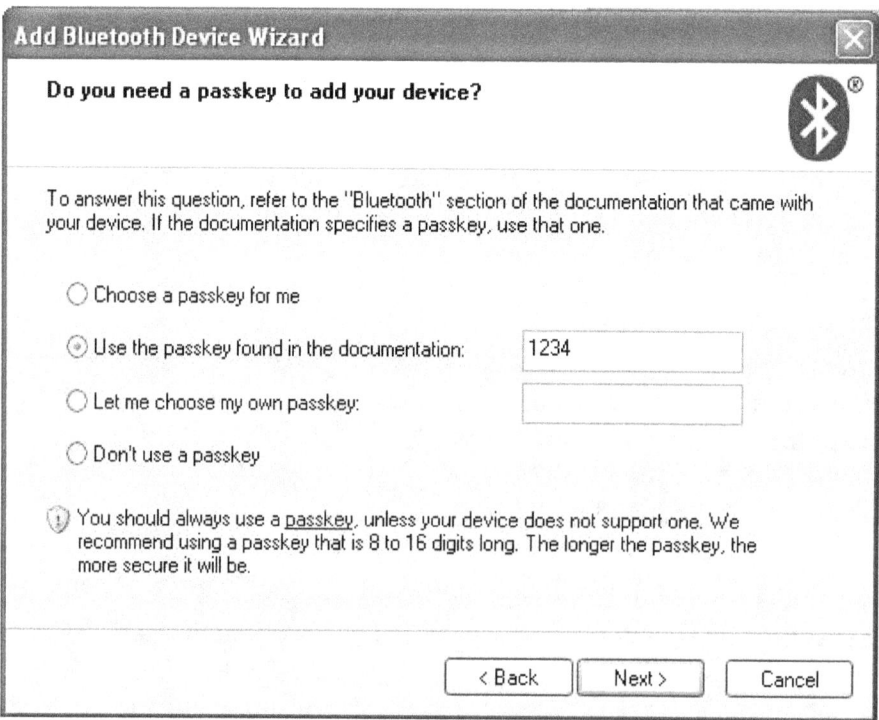

If the Bluetooth devices connect and everything works, the red light on the Bluetooth receiver will stop blinking and stay constantly lit. You should also get a message on the computer saying what serial port(s) have now been assigned to the Bluetooth device. It will look something like what you see in the next picture.

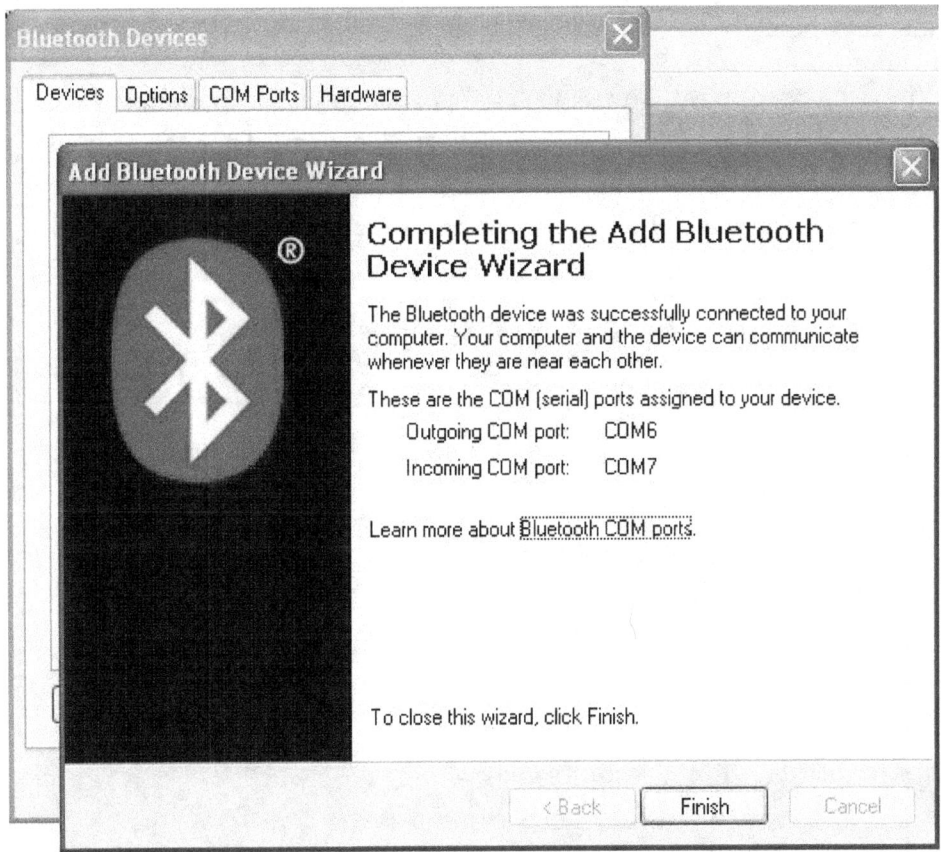

Next you need to load the serial sign sketch into your Arduino. Then connect the "TXD" pin of the Bluetooth receiver to the Arduino data pin D0 and the "RXD" pin to the Arduino pin D1. Note that connecting the Bluetooth adapter will block you from uploading any additional sketches until it is disconnected.

Next load a terminal program on the PC such as "HyperTerminal" and set it up for the first communications port that was assigned to your Bluetooth device, and "Serial," "9600 baud," "N," "8," and "1". Those settings should be the default settings. Save this instance of HyperTerminal as "Bluetooth" for future use.

At this point you should be able to type on your PC keyboard and see it on the LED sign. Congratulations, you have mastered connecting a PC to an Arduino via Bluetooth. I could not find any step by step instructions to do this anywhere!

You can check on the Bluetooth serial ports at any time by right clicking on "My Computer" and selecting "Device Manager". The screen should then show Bluetooth stuff in three locations like in the next picture. The Bluetooth adapter should come up under "Bluetooth Radios", "Network Adapters" and "Ports".

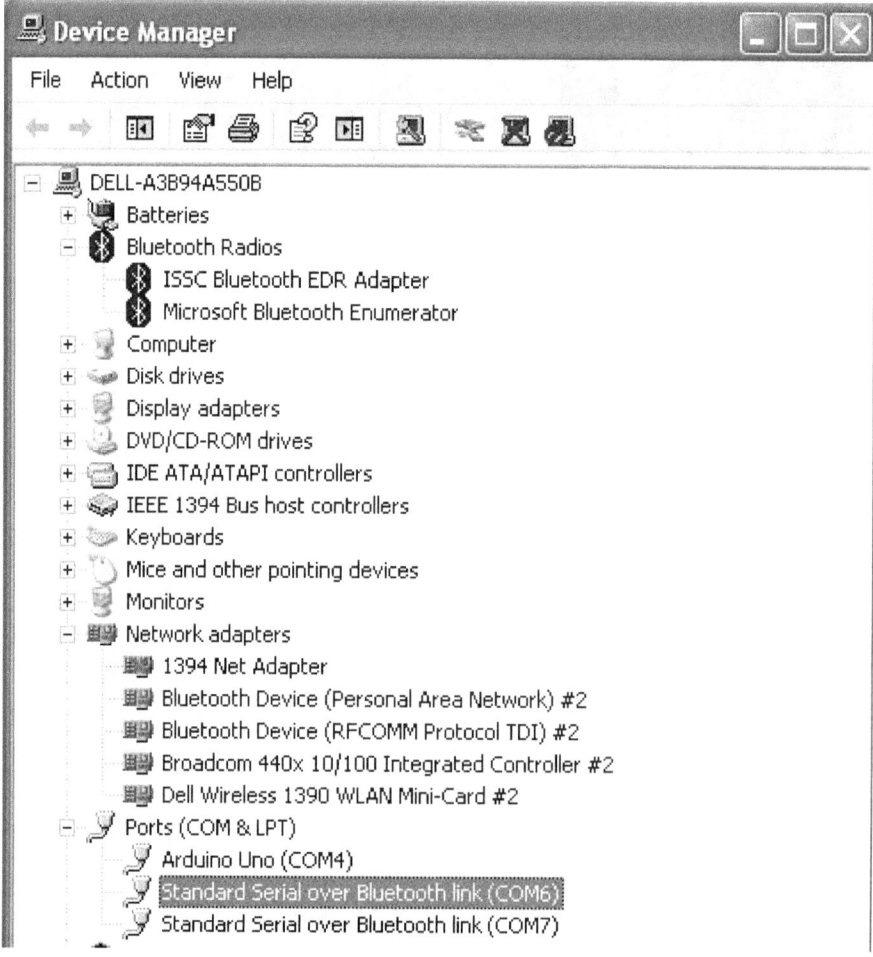

Up next is a picture showing how to connect the Bluetooth receiver to the Arduino. The Bluetooth module runs on 3.3 volts internally. Do not connect D1 directly to the Arduino, unless you have an adapter, use a 1K resistor in series, or simply do not use it. D1 is not needed to send commands from the PC to the Arduino. It is only needed to send responses back to the computer.

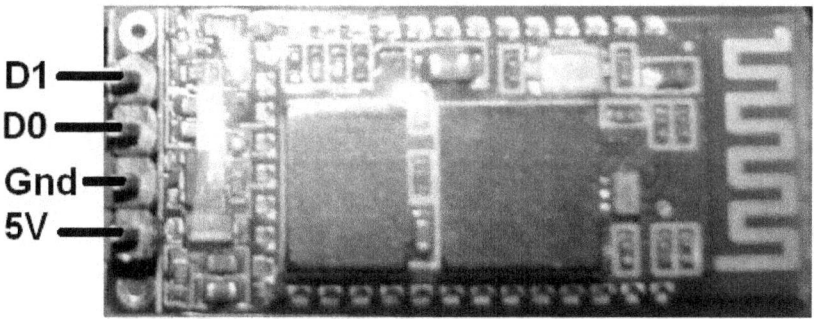

Please note that the Bluetooth connector is 6 pins but you only need to use four of them. The outermost two pins are not used. The top four Bluetooth pins line up as VCC, GND, TXD and RXD.

Using an Android phone can be a little difficult as well. Download and install "Serial Bluetooth Terminal". Start the ap and select the three bars in the upper left corner of the ap. Select "Devices" click on the "Gear" icon and select "Search for devices". Select the "HC-06" and "Pair the Device". Then enter the password of "1234". Hopefully the light on the receiver will then blink steady and the phone will say that they are paired.

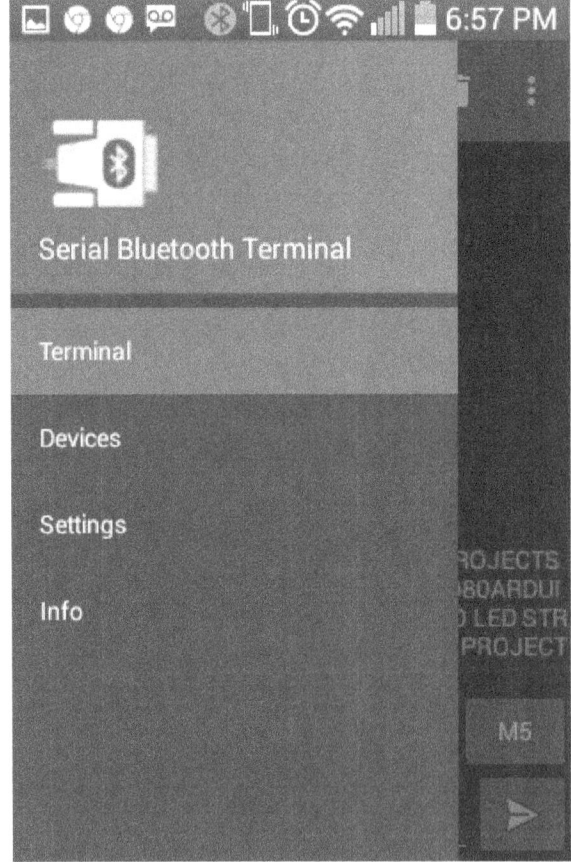

Hold down M1 and then type in a message for M1. You can type in five standard messages by using M1 to M5. You will need to send the messages before the fireworks or the entire message might not make it through to the sign in time. The message must always start with three numbers like "888". The first number is red, second number green and the last number is blue. The phone will repeat the last message for as long as it is connected to the sign.

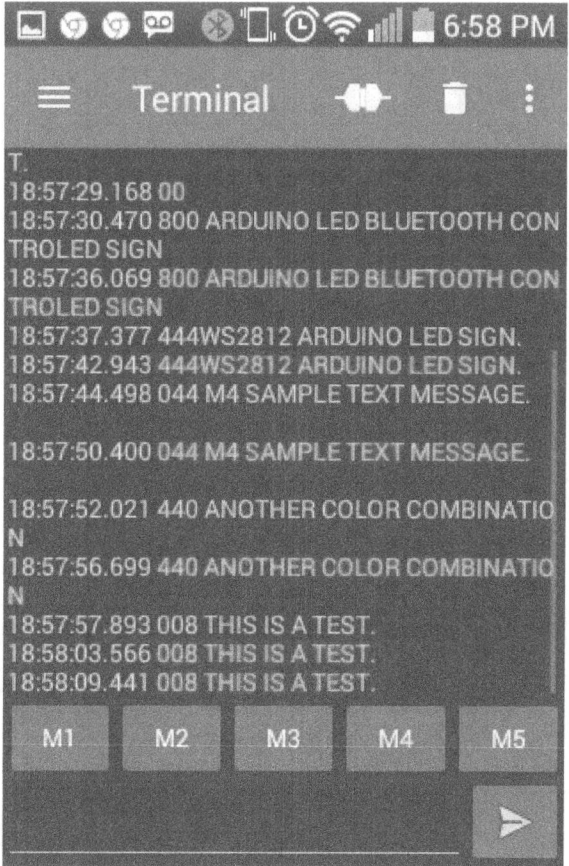

RF Wireless USB Adapter

Another option for wireless operation is to use wireless USB. This takes three parts, a USB to TTL adapter, and two bidirectional transceiver modules. I used my collection of parts to complete the plug in adapters.

Here is the wireless USB adapter that makes it easy to add remote control to your projects. I made an adapter to connect the USB to TTL adapter and the transmitter adapter together. Then I glued them into the adapter.

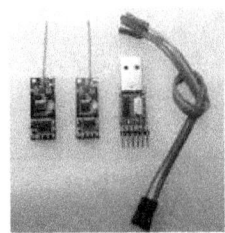

This is a picture of the assembled wireless USB transmitter adapter.

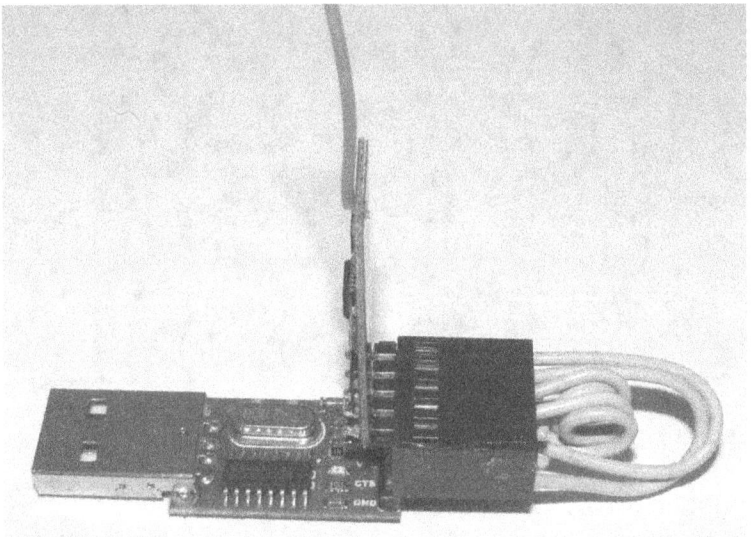

To make the adapter I used a dual six pin header, although you could also glue two six pin headers together. This next picture shows how to wire the two adapters together. You might note that you are connecting TX to the RX and vice versa. Once it is wired up and working you might want to cover the adapter connector with a removable flexible glue.

Next you can also need to make an adapter for the receiver. For this adapter you will need two five pin female headers.

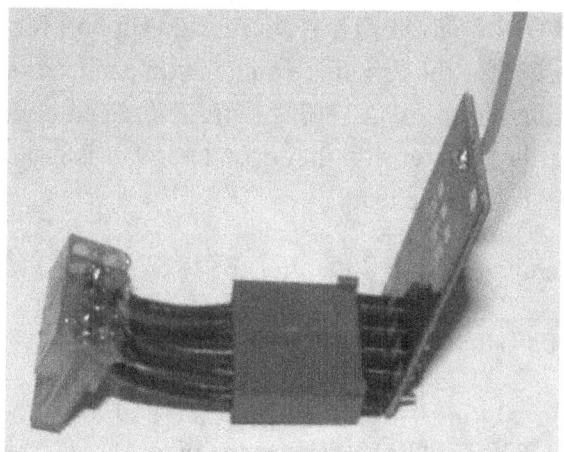

The wiring for this adapter is easier. The right two wires are direct and the next two - TX and RX must be crossed. The receiver adapter plugs into the Bluetooth jack or cable - the pins are now the same as the Bluetooth. You could also directly cable the receiver to the Arduino and skip making an adapter. However if you do that remember that power pins are on one side of the Arduino, D0 and D1 are on the other side.

The software had to be changed to add support for using port B and to not use bits D0 and D1 of port D. Then we can add support for serial communications. Also the first three numbers in each text string has to be three numbers ranging from 0 to 8 that represent the color combination. The first number is red intensity, the second number is green intensity and the third number is blue intensity.

```
// Version for sending data to 8 Parallel WS2812 strings
// by: Bob Davis
// Shifted to Ports B and D to allow serial communications

// PORTD is Digital Pins 0-7 on the Uno change for other boards.
#define PIXEL_PORT  PORTD  // Port the pixels are connected to
#define PIXEL_DDR   DDRD   // Port the pixels are connected to
#define PIXEL_PORTB PORTB  // Port the pixels are connected to
#define PIXEL_DDRB  DDRB   // Port the pixels are connected to
int textb=128; // brightness of text
int maxled=180;
// Default line of text in Green.
String text="080 ARDUINO LED STRIP PROJECTS    ";
```

```
// Send the next set of 8 WS2812B encoded bits to the 8 pins.
// The delay timing is for an Arduino UNO.
void sendBitX8( byte bits ) {
  PORTD= 0xFF;  // turn on
  PORTB= 0xFF;  // turn on
  PORTD= 0xFF;  // delay
  PORTD= 0xFF;  // delay
  PORTD= 0xFF;  // delay (add more for faster processors)
  PORTD= bits;  // send data
  PORTB= bits;  // send data
  PORTD= bits;  // delay
  PORTD= bits;  // delay
  PORTD= bits;  // delay
  PORTD= 0x00;  // Turn off;
  PORTB= 0x00;  // Turn off;
}

// Set default color for letters
int red=1; int green=textb; int blue=1;

void sendPixelRow( byte row ) {
  // Send the bit 8 times down every row, each pixel is 8 bits/color
  int mask = 0x01;  // shifting mask to determine bit status
    for (int bit=8; bit>0; bit--){
      if (green & (mask<<bit))sendBitX8( row ); else sendBitX8( 0x00 ); }
    for (int bit=8; bit>0; bit--){
      if (red & (mask<<bit))sendBitX8( row ); else sendBitX8( 0x00 ); }
    for (int bit=8; bit>0; bit--){
      if (blue & (mask<<bit))sendBitX8( row ); else sendBitX8( 0x00 ); }
}

// This font from http://sunge.awardspace.com/glcd-sd/node4.html
byte font[][6] = {
0x00,0x00,0x00,0x00,0x00,0x00, // ascii 32
0x00,0x00,0xfa,0x00,0x00,0x00, // !
0x00,0xe0,0x00,0xe0,0x00,0x00, // "
0x28,0xfe,0x28,0xfe,0x28,0x00, // #
0x24,0x54,0xfe,0x54,0x48,0x00, // $
0xc4,0xc8,0x10,0x26,0x46,0x00, // %
0x6c,0x92,0xaa,0x44,0x0a,0x00, // &
0x00,0xa0,0xc0,0x00,0x00,0x00, // '
```

```
0x00,0x38,0x44,0x82,0x00,0x00, // (
0x00,0x82,0x44,0x38,0x00,0x00, // )
0x10,0x54,0x38,0x54,0x10,0x00, // *
0x10,0x10,0x7c,0x10,0x10,0x00, // +
0x00,0x0a,0x0c,0x00,0x00,0x00, // ,
0x10,0x10,0x10,0x10,0x10,0x00, // -
0x00,0x06,0x06,0x00,0x00,0x00, // .
0x04,0x08,0x10,0x20,0x40,0x00, // /
0x7c,0x8a,0x92,0xa2,0x7c,0x00, // 0
0x00,0x42,0xfe,0x02,0x00,0x00, // 1
0x42,0x86,0x8a,0x92,0x62,0x00, // 2
0x84,0x82,0xa2,0xd2,0x8c,0x00, // 3
0x18,0x28,0x48,0xfe,0x08,0x00, // 4
0xe4,0xa2,0xa2,0xa2,0x9c,0x00, // 5
0x3c,0x52,0x92,0x92,0x0c,0x00, // 6
0x80,0x8e,0x90,0xa0,0xc0,0x00, // 7
0x6c,0x92,0x92,0x92,0x6c,0x00, // 8
0x60,0x92,0x92,0x94,0x78,0x00, // 9
0x00,0x6c,0x6c,0x00,0x00,0x00, // :
0x00,0x6a,0x6c,0x00,0x00,0x00, // ;
0x00,0x10,0x28,0x44,0x82,0x00, // <
0x28,0x28,0x28,0x28,0x28,0x00, // =
0x82,0x44,0x28,0x10,0x00,0x00, // >
0x40,0x80,0x8a,0x90,0x60,0x00, // ?
0x4c,0x92,0x9e,0x82,0x7c,0x00, // @
0x7e,0x90,0x90,0x90,0x7e,0x00, // A
0xfe,0x92,0x92,0x92,0x6c,0x00, // B
0x7c,0x82,0x82,0x82,0x44,0x00, // C
0xfe,0x82,0x82,0x44,0x38,0x00, // D
0xfe,0x92,0x92,0x92,0x82,0x00, // E
0xfe,0x90,0x90,0x80,0x80,0x00, // F
0x7c,0x82,0x82,0x8a,0x4c,0x00, // G
0xfe,0x10,0x10,0x10,0xfe,0x00, // H
0x00,0x82,0xfe,0x82,0x00,0x00, // I
0x04,0x02,0x82,0xfc,0x80,0x00, // J
0xfe,0x10,0x28,0x44,0x82,0x00, // K
0xfe,0x02,0x02,0x02,0x02,0x00, // L
0xfe,0x40,0x20,0x40,0xfe,0x00, // M
0xfe,0x20,0x10,0x08,0xfe,0x00, // N
0x7c,0x82,0x82,0x82,0x7c,0x00, // O
0xfe,0x90,0x90,0x90,0x60,0x00, // P
```

```
0x7c,0x82,0x8a,0x84,0x7a,0x00, // Q
0xfe,0x90,0x98,0x94,0x62,0x00, // R
0x62,0x92,0x92,0x92,0x8c,0x00, // S
0x80,0x80,0xfe,0x80,0x80,0x00, // T
0xfc,0x02,0x02,0x02,0xfc,0x00, // U
0xf8,0x04,0x02,0x04,0xf8,0x00, // V
0xfe,0x04,0x18,0x04,0xfe,0x00, // W
0xc6,0x28,0x10,0x28,0xc6,0x00, // X
0xc0,0x20,0x1e,0x20,0xc0,0x00, // Y
0x86,0x8a,0x92,0xa2,0xc2,0x00, // Z
};

void setup() {
  Serial.begin(9600);    // opens serial port, sets data rate to 9600 bps
  PIXEL_DDR = 0xF0;      // Set pins to output Cannot use D0 and D1!
  PIXEL_DDRB = 0xFF;     // Set pins to output
}
void loop() {
// Serial.readBytesUntil(stopc, text, 30);
  if (Serial.available() > 0) text=Serial.readString();
  Serial.print(text);
  red=0; green=textb; blue=0;
  red=text[0]*32;
  green=text[1]*32;
  blue=text[2]*32;

  cli();              // No time for interruptions!
  for (int l=3; l<33; l++){    // 30 is number of characters to send
    // Using a loop can be too slow with the character lookups.
    sendPixelRow(font[text[l]-32][0]); // -32 is to get to correct ascii
    sendPixelRow(font[text[l]-32][1]); // -32 is to get to correct ascii
    sendPixelRow(font[text[l]-32][2]); // -32 is to get to correct ascii
    sendPixelRow(font[text[l]-32][3]); // -32 is to get to correct ascii
    sendPixelRow(font[text[l]-32][4]); // -32 is to get to correct ascii
    sendPixelRow(font[text[l]-32][5]); // -32 is to get to correct ascii
  }
  sei();
  delay(4000);           // Wait to display results

  cli();              // No time for interruptions!
  for (int l=0; l<maxled; l++){    // 30 is number of characters to clear
```

```
    sendPixelRow(0x00);      // clear the screen
  }
  sei();

// FIREWORKS
  cli();                     // No time for interruptions!
  red=textb;  green=textb;  blue=textb;
  for (int cycle=0; cycle<100; cycle++){
    int rabit=random(7);
    byte ls=0x01;
    byte sbit=ls << rabit;      // Set up random bit
    int rano=random(maxled);       // Set Random location
    for (int l=0; l<maxled; l++){  // 180 is number of bytes to send
      if (l==rano) sendPixelRow(sbit); else sendPixelRow(0x00);
    }
    delay(20);
  }
  sei();
  return;
}
```

Chapter 6

Interfacing to 12 LED Strips

If you use both ports D and B at the same time you can send 12 or even 14 bits to the LED sign at a time. This makes it possible to use larger, neater fonts. But first we have to create those larger fonts. Most fonts are oriented horizontal or "raster" based for a LCD screen. That is to say the numbers represent bits from left to right. For the LED sign we will need fonts with the bits recorded vertically from top to bottom.

Here is a picture of a typical 5x7 font with the values both top to bottom and right to left. As you can see they are totally different numbers. So for my 12 line font, a new font had to be created. I printed out a font from a font.c library and then typed in the numbers as they are read from the different direction to create my new font.

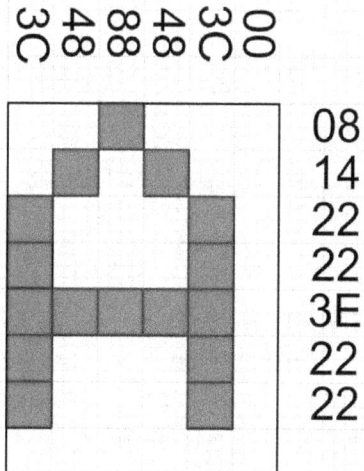

The next picture shows a typical 12x8 font. The bottom two rows are reserved for lower case letters that extend below the normal line. On the right are the normal numbers but across the top are the new numbers that you need for a parallel LED sign. Note that then numbers are now three digits long.

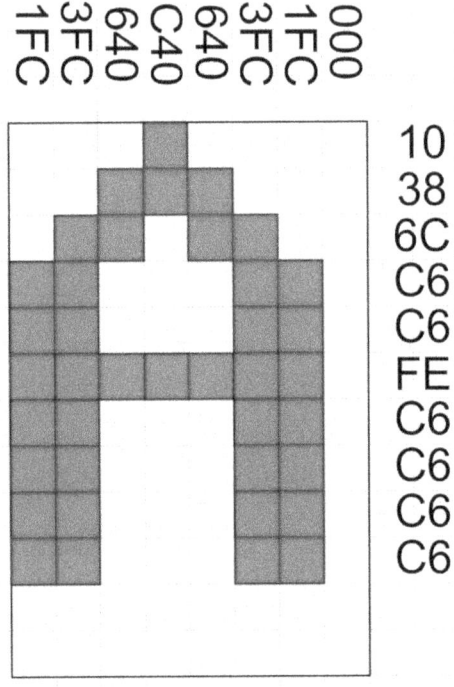

This next picture is what the 12 Strip LED sign looks like with the new bigger font.

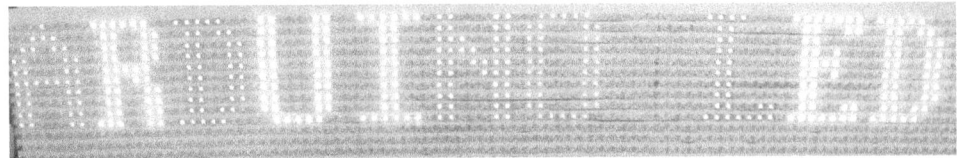

This time the pin connections to the Arduino are in numerical order as can be seen in the next schematic diagram. Basically you will need to make another four line wiring harness to connect to the additional four of the twelve lines from the Arduino to the sign. The connection numbers are referring to the Data pin designations on the Arduino. Pins D8 to D11 are Port B in the software.

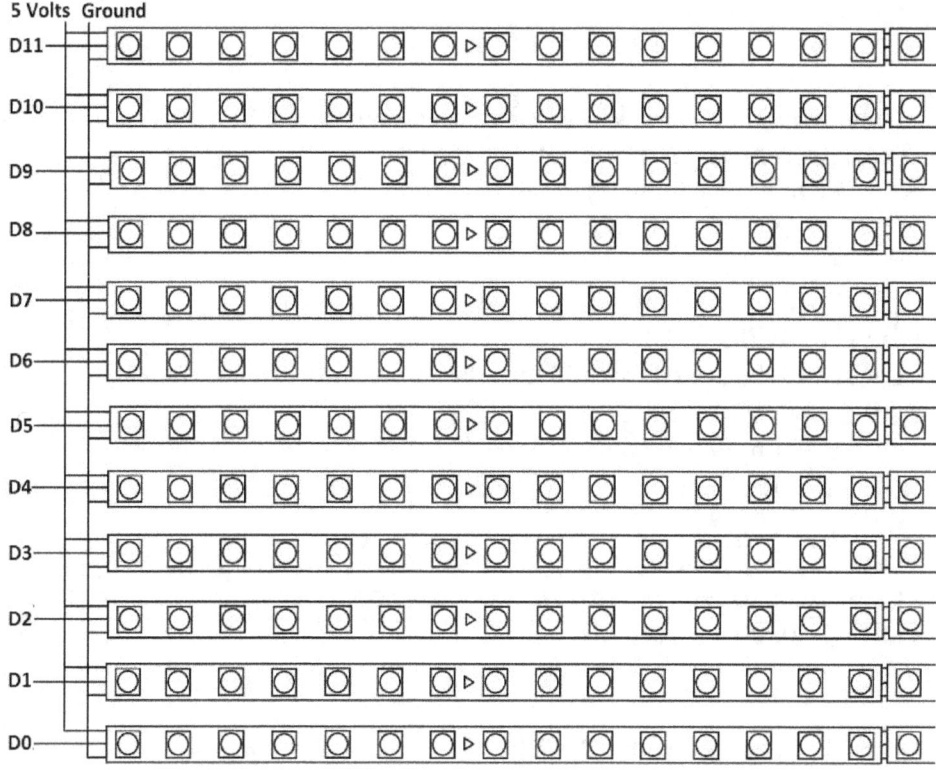

This software reflects the new 12x8 font. I did not type in all of the lower case letters but they would use the lower two bits more than the upper case letters do. Basically you could also use this for a 10x8 font. The 16x90 line sign used to hold two rows of 15 letters each, but now it only holds 11 of the larger letters.

```
// Version for sending data to 12 Parallel WS2812 strings
// by: BOB Davis
// Removed Assembler and simplified the code

// PORTD is Digital Pins 0-7 on the Uno change for other boards.
#define PIXEL_PORT  PORTD  // Port the pixels are connected to
#define PIXEL_DDR   DDRD   // Port the pixels are connected to
// PORT B is D8-D13
#define PIXEL_PORTB PORTB  // Port the pixels are connected to
#define PIXEL_DDRB  DDRB   // Port the pixels are connected to
int textb=128;  // brightness of text
int maxled=90;
String text1="ARDUINO LED STRIP PROJECTS    ";
```

```
// Set default color for letters
int red=textb; int green=textb; int blue=textb;

// Send the next set of 8 WS2812B encoded bits to the 8 pins.
// The delay timing is for an Arduino UNO.
void sendBitX8( byte bits, byte bith ) {
  PORTD= 0xFF;  // turn on
  PORTB= 0xFF;  // turn on
  PORTD= 0xFF;  // delay
  PORTD= 0xFF;  // delay
  PORTD= 0xFF;  // delay (add more for faster processors)
  PORTD= bits;  // send data
  PORTB= bith;  // send data
  PORTD= bits;  // delay
  PORTD= bits;  // delay
  PORTD= bits;  // delay
  PORTD= 0x00;  // Turn off;
  PORTB= 0x00;  // Turn off;
}

void sendPixelRow( word row ) {
  // separate out upper and lower bytes from word
  byte rowh = row >> 8;
  // Send the bit 8 times down every row, each pixel is 8 bits/color
  int mask = 0x01;  // shifting mask to determine bit status
    for (int bit=8; bit>0; bit--){
     if (green & (mask<<bit))sendBitX8( row, rowh ); else sendBitX8( 0x0000, 0x0000 ); }
    for (int bit=8; bit>0; bit--){
     if (red & (mask<<bit))sendBitX8( row, rowh ); else sendBitX8( 0x0000, 0x0000 ); }
    for (int bit=8; bit>0; bit--){
     if (blue & (mask<<bit))sendBitX8( row, rowh ); else sendBitX8( 0x0000, 0x0000 ); }
  }


// Hand typed 8x12 Font rotated to vertical values
// Based on Michael J. Karas of Carousel Design Solutions
word font[][8] = {
0x00,0x00,0x00,0x00,0x00,0x00,0x00,0x00, // ascii 32
```

```
0x000,0x000,0x780,0xFEC,0xFEC,0x780,0x000,0x000, // !
0x000,0xE00,0xF00,0x000,0x000,0xF00,0xE00,0x000, // "
0x110,0x3FC,0x3FC,0x110,0x3FC,0x3FC,0x110,0x000, // #
0x31C,0x7C6,0x441,0xFFF,0xC43,0x67E,0x33C,0x000, // $
0xE18,0xA30,0xE60,0x0C0,0x19C,0x314,0x61C,0x000, // %
0x678,0xFFC,0x984,0x9C4,0xF78,0x63C,0x068,0x000, // &
0x000,0x000,0x100,0xF00,0xE00,0x000,0x000,0x000, // '
0x000,0x000,0x3F0,0x7F8,0xC0C,0x804,0x000,0x000, // (
0x000,0x000,0x804,0xC0C,0x7F8,0x3F0,0x000,0x000, // )
0x040,0x150,0x1F0,0x0E0,0x1F0,0x150,0x040,0x000, // *
0x0C0,0x0C0,0x0C0,0x3F0,0x3F0,0x0C0,0x0C0,0x0C0, // +
0x000,0x000,0x001,0x00F,0x00E,0x000,0x000,0x000, // ,
0x040,0x040,0x040,0x040,0x040,0x040,0x040,0x000, // -
0x000,0x000,0x00C,0x00C,0x000,0x000,0x000,0x000, // .
0x006,0x00C,0x018,0x030,0x060,0x0C0,0x180,0x300, // /
0x3F0,0x7F8,0xC0C,0x8C4,0xC0C,0x7F8,0x3F0,0x000, // 0
0x000,0x404,0x604,0xFFC,0xFFC,0x004,0x004,0x000, // 1
0x41C,0xC3C,0x864,0x8C4,0x984,0xF0C,0x60C,0x000, // 2
0x408,0xC0C,0x804,0x884,0x884,0xFFC,0x778,0x000, // 3
0x0E0,0x1E0,0x320,0x624,0xFFC,0xFFC,0x024,0x000, // 4
0xF88,0xF8C,0x884,0x884,0x884,0x8FC,0x878,0x000, // 5
0x3F8,0x7FC,0xC84,0x884,0x884,0x8FC,0x078,0x000, // 6
0xC00,0xC00,0x87C,0x8FC,0x980,0xF00,0xE00,0x000, // 7
0x778,0xFFC,0x884,0x884,0x884,0xFFC,0x778,0x000, // 8
0x700,0xF84,0x884,0x884,0x88C,0xFF8,0x7F0,0x000, // 9
0x000,0x000,0x000,0x318,0x318,0x000,0x000,0x000, // :
0x000,0x000,0x002,0x18E,0x18C,0x000,0x000,0x000, // ;
0x000,0x040,0x0E0,0x1B0,0x318,0x60C,0x404,0x000, // <
0x000,0x120,0x120,0x120,0x120,0x120,0x120,0x000, // =
0x000,0x404,0x60C,0x318,0x1B0,0x0E0,0x040,0x000, // >
0x400,0xC00,0x800,0x86C,0x8EC,0xF80,0x700,0x000, // ?
0x7F8,0xFFC,0x804,0x864,0x8E4,0xFE4,0x7C0,0x000, // @
0x1FC,0x3FC,0x640,0xC40,0x640,0x3FC,0x1FC,0x000, // A
0x804,0xFFC,0xFFC,0x884,0x884,0xFFC,0x778,0x000, // B
0x3F0,0x7F8,0xC0C,0x804,0x804,0xC0C,0x618,0x000, // C
0x804,0xFFC,0xFFC,0x804,0xC0C,0x7F8,0x3F0,0x000, // D
0x804,0xFFC,0xFFC,0x884,0x9C4,0xC0C,0xC0C,0x000, // E
0x804,0xFFC,0xFFC,0x884,0x9C0,0xC00,0xC00,0x000, // F
0x3F0,0x7F8,0xC0C,0x804,0x844,0xC78,0x67C,0x000, // G
0xFFC,0xFFC,0x080,0x080,0x080,0xFFC,0xFFC,0x000, // H
0x000,0x804,0x804,0xFFC,0xFFC,0x804,0x804,0x000, // I
```

```
0x018,0x01C,0x004,0x804,0xFFC,0xFF8,0x800,0x000, // J
0x804,0xFFC,0xFFC,0x1C0,0x360,0xE3C,0xC1C,0x000, // K
0x804,0xFFC,0xFFC,0x804,0x004,0x00C,0x00C,0x000, // L
0xFFC,0xFFC,0x700,0x380,0x700,0xFFC,0xFFC,0x000, // M
0xFFC,0xFFC,0x300,0x180,0x0C0,0xFFC,0xFFC,0x000, // N
0x7F8,0xFFC,0x804,0x804,0x804,0xFFC,0x7F8,0x000, // O
0x804,0xFFC,0xFFC,0x844,0x840,0xFC0,0x780,0x000, // P
0x7F8,0xFFC,0x808,0x80C,0x806,0xFFF,0x7F9,0x000, // Q
0x804,0xFFC,0xFFC,0x880,0x8C0,0xFFC,0x73C,0x000, // R
0x608,0xF0C,0x984,0x884,0x8C4,0xC7C,0x438,0x000, // S
0xC00,0xC00,0xC04,0xFFC,0xFFC,0xC04,0xC00,0xC00, // T
0xFF8,0xFFC,0x004,0x004,0x004,0xFFC,0xFF8,0x000, // U
0xFE0,0xFF0,0x018,0x00C,0x018,0xFF0,0xFE0,0x000, // V
0xFF0,0xFFC,0x01C,0x070,0x01C,0xFFC,0xFF0,0x000, // W
0xC0C,0xE1C,0x330,0x1E0,0x1E0,0x330,0xE1C,0xC0C, // X
0xE00,0xF00,0x184,0x0FC,0x0FC,0x184,0xF00,0xE00, // Y
0xC1C,0xC3C,0x864,0x8C4,0x984,0xB04,0xE0C,0x000, // Z
};


void setup() {
  PIXEL_DDR = 0xFF;    // Set all row pins to output
  PIXEL_DDRB = 0xFF;   // Set pins to output
}
void loop() {
// 256 SHADE COLOR TEST
  cli();              // No time for interruptions!
  for (int l=0; l<(maxled/4); l++){ // max number lines to send
    red=l+2; green=0; blue=0;
    sendPixelRow(0xFFF);      // Light red
  }
  for (int l=0; l<(maxled/4); l++){ // max number lines to send
    red=0; green=l+2; blue=0;
    sendPixelRow(0xFFF);      // Light green
  }
  for (int l=0; l<(maxled/4); l++){ // max number lines to send
    red=0; green=0; blue=l+2;
    sendPixelRow(0xFFF);      // Light blue
  }
  for (int l=0; l<(maxled/4); l++){ // max number lines to send
    red=l+2; green=l+2; blue=l+2;
```

```c
    sendPixelRow(0xFFF);      // Light white
  }
  sei();
  delay(3000);                // Wait to display results

  red=textb; green=0; blue=0;
  cli();                      // No time for interruptions!
  for (int l=0; l<30; l++){   // 30 is number of characters to send
    switch (l){               // case is faster than if's
      case 0: red=textb; green=0; blue=0; break; // red
      case 1: red=0; green=textb; blue=0; break;  // green
      case 2: red=0; green=0; blue=textb; break;  // blue
      case 3: red=textb; green=textb; blue=0; break; //yellow
      case 4: red=0; green=textb; blue=textb; break; // cyan
      case 5: red=textb; green=0; blue=textb; break; // purple
      case 6: red=textb; green=0; blue=0; break;  // red
      case 7: red=0; green=textb; blue=0; break;  // green
      case 8: red=0; green=0; blue=textb; break;  // blue
      case 9: red=textb; green=textb; blue=0; break; //yellow
      case 10: red=0; green=textb; blue=textb; break; // cyan
      case 11: red=textb; green=0; blue=textb; break; // purple
    }
    // Using a loop is too slow with the character lookups.
    sendPixelRow(font[text1[l]-32][0]); // -32 is to get to correct ascii
    sendPixelRow(font[text1[l]-32][1]); // -32 is to get to correct ascii
    sendPixelRow(font[text1[l]-32][2]); // -32 is to get to correct ascii
    sendPixelRow(font[text1[l]-32][3]); // -32 is to get to correct ascii
    sendPixelRow(font[text1[l]-32][4]); // -32 is to get to correct ascii
    sendPixelRow(font[text1[l]-32][5]); // -32 is to get to correct ascii
    sendPixelRow(font[text1[l]-32][6]); // -32 is to get to correct ascii
    sendPixelRow(font[text1[l]-32][7]); // -32 is to get to correct ascii
  }
  sei();
  delay(8000);                // Wait to display results
  return;
}
```

Chapter 7

Displaying Graphics

Someone saw my sign and questioned "Does it do graphics?" Well yes it can. The coding can be a bit tricky. Like with the fonts the graphics has to be turned sideways. First let me explain BMP graphics. There are two common BMP formats. One is called 565 and the other is RGB. Most of the other graphics formats use compression and that gets beyond simple coding.

In BMP 565 there are two bytes containing five bits of red, six bits of green and five bits of blue. It is 33% smaller that full RGB at 16 bits per pixel.

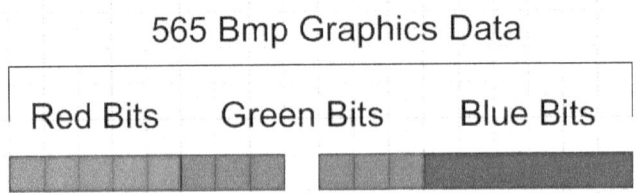

RGB has three bytes one of red, one of green and one of blue. That is 24 bits per pixel. To turn RGB sideways we need to take samples look at the MSB and use that to set bits in an array that will be sent to the sign. An easier way is to just manually encode the graphics for the sign to begin with.

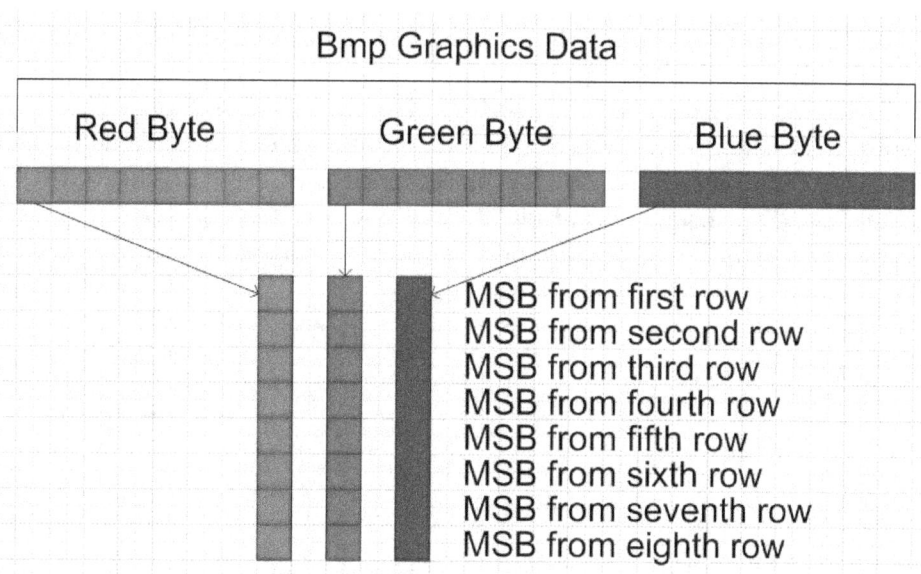

This flag was created manually by typing in the color values. There are three bytes per vertical column. They are red, green, and blue. For instance "0xFF,0x55,0x55" produces red and white stripes. the red value is full on but the green and blue values turn off for every other line.

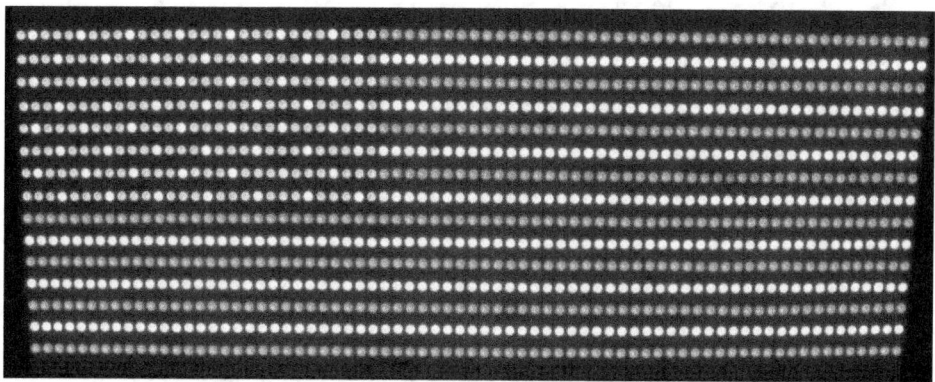

Here is a program that uses three bytes per color and column position and it produces an American flag on a sign with two rows of 70 to 90 columns per row. The graphics was hand coded as vertical R G B columns. It adds two extra states and starts with three extra colonies, but it is a close representation of the American flag.

```
// Version for sending data to 8 Parallel WS2812 strings
// by: BOB Davis
```

// Graphics version with Manually entered US Flag

```
// PORTD is Digital Pins 0-7 on the Uno change for other boards.
#define PIXEL_PORT  PORTD  // Port the pixels are connected to
#define PIXEL_DDR   DDRD   // Port the pixels are connected to
int textb=128;  // brightness of text
int maxled=180;  // maximum number of LED's insign
//byte sarray[144*3]; // three times max led
int imgwidth=72;  // image number of pixels wide
byte mbyte;
#include <avr/pgmspace.h>

// Set default color for letters
int red=10; int green=10; int blue=10;

// Send the next set of 8 WS2812B encoded bits to the 8 pins.
// The delay timing is for an Arduino UNO.
void sendBitX8( byte bits ) {
  PORTD= 0xFF;  // turn on
  PORTD= 0xFF;  // delay
  PORTD= 0xFF;  // delay
  PORTD= 0xFF;  // delay
  PORTD= 0xFF;  // delay (add more for faster processors)
  PORTD= bits;  // send data
  PORTD= bits;  // delay
  PORTD= bits;  // delay
  PORTD= bits;  // delay
  PORTD= bits;  // delay
  PORTD= 0x00;  // Turn off;
}

int mask = 0x01;  // will shift mask to determine bit status
void sendPixelRow( byte mred, byte mgreen, byte mblue ) {
  // Send the bit 8 times down every row, each pixel is 8 bits/color
    for (int mbit=8; mbit>0; mbit--){
     if (green & mask<<mbit)sendBitX8( mgreen ); else sendBitX8( 0x00
);}
    for (int mbit=8; mbit>0; mbit--){
     if (red & mask<<mbit)sendBitX8( mred ); else sendBitX8( 0x00 ); }
    for (int mbit=8; mbit>0; mbit--){
     if (blue & mask<<mbit)sendBitX8( mblue ); else sendBitX8( 0x00 ); }
```

}

```
static byte sarray[] = {
// Left corner
0x00,0x00,0xFF, 0xAA,0xAA,0xFF, 0x00,0x00,0xFF, 0x55,0x55,0xFF,
0x00,0x00,0xFF, 0xAA,0xAA,0xFF, 0x00,0x00,0xFF, 0x55,0x55,0xFF,
0x00,0x00,0xFF, 0xAA,0xAA,0xFF, 0x00,0x00,0xFF, 0x55,0x55,0xFF,
0x00,0x00,0xFF, 0xAA,0xAA,0xFF, 0x00,0x00,0xFF, 0x55,0x55,0xFF,
0x00,0x00,0xFF, 0xAA,0xAA,0xFF, 0x00,0x00,0xFF, 0x55,0x55,0xFF,
0x00,0x00,0xFF, 0xAA,0xAA,0xFF, 0x00,0x00,0xFF, 0x55,0x55,0xFF,
0x00,0x00,0xFF, 0xAA,0xAA,0xFF, 0x00,0x00,0xFF,
// Right Side + Bottom
0xFF,0x55,0x55, 0xFF,0x55,0x55, 0xFF,0x55,0x55, 0xFF,0x55,0x55,
0xFF,0x55,0x55, 0xFF,0x55,0x55, 0xFF,0x55,0x55, 0xFF,0x55,0x55,
0xFF,0x55,0x55, 0xFF,0x55,0x55, 0xFF,0x55,0x55, 0xFF,0x55,0x55,
0xFF,0x55,0x55, 0xFF,0x55,0x55, 0xFF,0x55,0x55, 0xFF,0x55,0x55,
0xFF,0x55,0x55, 0xFF,0x55,0x55, 0xFF,0x55,0x55, 0xFF,0x55,0x55,
0xFF,0x55,0x55, 0xFF,0x55,0x55, 0xFF,0x55,0x55, 0xFF,0x55,0x55,
0xFF,0x55,0x55, 0xFF,0x55,0x55, 0xFF,0x55,0x55, 0xFF,0x55,0x55,
0xFF,0x55,0x55, 0xFF,0x55,0x55, 0xFF,0x55,0x55, 0xFF,0x55,0x55,
0xFF,0x55,0x55, 0xFF,0x55,0x55, 0xFF,0x55,0x55, 0xFF,0x55,0x55,
0xFF,0x55,0x55, 0xFF,0x55,0x55, 0xFF,0x55,0x55, 0xFF,0x55,0x55,
0xFF,0x55,0x55, 0xFF,0x55,0x55, 0xFF,0x55,0x55, 0xFF,0x55,0x55,
0xFF,0x55,0x55, 0xFF,0x55,0x55, 0xFF,0x55,0x55, 0xFF,0x55,0x55,
0xFF,0x55,0x55, 0xFF,0x55,0x55, 0xFF,0x55,0x55, 0xFF,0x55,0x55,
0xFF,0x55,0x55, 0xFF,0x55,0x55, 0xFF,0x55,0x55, 0xFF,0x55,0x55,
0xFF,0x55,0x55, 0xFF,0x55,0x55, 0xFF,0x55,0x55, 0xFF,0x55,0x55,
0xFF,0x55,0x55, 0xFF,0x55,0x55, 0xFF,0x55,0x55, 0xFF,0x55,0x55,
0xFF,0x55,0x55, 0xFF,0x55,0x55, 0xFF,0x55,0x55, 0xFF,0x55,0x55,
0xFF,0x55,0x55, 0xFF,0x55,0x55, 0xFF,0x55,0x55, 0xFF,0x55,0x55,
0xFF,0x55,0x55, 0xFF,0x55,0x55, 0xFF,0x55,0x55, 0xFF,0x55,0x55,
0xFF,0x55,0x55, 0xFF,0x55,0x55, 0xFF,0x55,0x55, 0xFF,0x55,0x55,
0xFF,0x55,0x55, 0xFF,0x55,0x55, 0xFF,0x55,0x55, 0xFF,0x55,0x55,
0xFF,0x55,0x55, 0xFF,0x55,0x55, 0xFF,0x55,0x55, 0xFF,0x55,0x55,
0xFF,0x55,0x55, 0xFF,0x55,0x55, 0xFF,0x55,0x55, 0xFF,0x55,0x55,
0xFF,0x55,0x55, 0xFF,0x55,0x55, 0xFF,0x55,0x55, 0xFF,0x55,0x55,
```

```
  0xFF,0x55,0x55, 0xFF,0x55,0x55, 0xFF,0x55,0x55, 0xFF,0x55,0x55,
  0xFF,0x55,0x55, 0xFF,0x55,0x55, 0xFF,0x55,0x55, 0xFF,0x55,0x55,
  0xFF,0x55,0x55, 0xFF,0x55,0x55, 0xFF,0x55,0x55, 0xFF,0x55,0x55,
  0xFF,0x55,0x55, 0xFF,0x55,0x55, 0xFF,0x55,0x55, 0xFF,0x55,0x55,
  0xFF,0x55,0x55, 0xFF,0x55,0x55, 0xFF,0x55,0x55, 0xFF,0x55,0x55,
  0xFF,0x55,0x55, 0xFF,0x55,0x55, 0xFF,0x55,0x55, 0xFF,0x55,0x55,
  0xFF,0x55,0x55, 0xFF,0x55,0x55, 0xFF,0x55,0x55, 0xFF,0x55,0x55,
  0xFF,0x55,0x55, 0xFF,0x55,0x55, 0xFF,0x55,0x55, 0xFF,0x55,0x55,
  0xFF,0x55,0x55, 0xFF,0x55,0x55, 0xFF,0x55,0x55, 0xFF,0x55,0x55,
  0xFF,0x55,0x55
};

void setup() {
  PIXEL_DDR = 0xff;   // Set all row pins to output
}
void loop() {
// 256 SHADE COLOR TEST
  int mqtr=maxled/4;  // doing the math several times is too slow
  int mhalf=maxled/2; // doing the math several times is too slow
  int m3qtr=maxled*.75; // doing the math several times is too slow
  cli();              // No time for interruptions!
  for (int l=0; l<(maxled); l++){ // max number lines to send
    red=l; green=0; blue=0;
    if (l >= mqtr)  { red=0; green=l-mqtr; blue=0;}
    if (l >= mhalf) { red=0; green=0; blue=l-mhalf;}
    if (l >= m3qtr) { red=l-m3qtr; green=l-m3qtr; blue=l-m3qtr;}
    sendPixelRow( 0xFF, 0xFF, 0xFF );     // Light bits
  }
  sei();
  delay(2000);              // Wait to display results

  red=textb; green=textb; blue=textb;
  int mmax=maxled*3;
  cli();                    // No time for interruptions!
  for (int l=0; l<mmax; l=l+3){    //  send array
    sendPixelRow(sarray[l], sarray[l+1], sarray[l+2] );
  }
  sei();
  delay(4000);              // Wait to display results
  return;
}
```

The other option is to use a BMP image and import it. To start with, you need to reduce the resolution to the width of your sign. I use IrfanView but you can also use Paint. Then you will need to use a program that removes the BMP header and inserts commas between all of the image bytes. I have used a program called "LCD Image Converter". It is free to download found on the Internet.

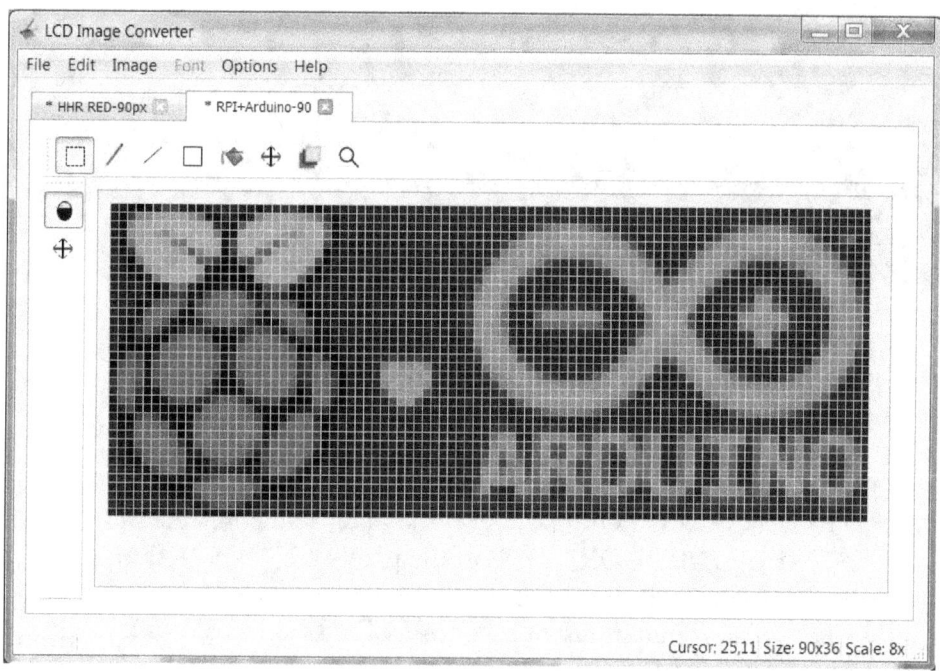

In LCD Image Converter you will need to go into "Options" and set it to "Color" and to three byte "RGB". Then select "File" and "Convert". The resulting output file will be too big to fit in the Arduino memory. You will need to add PROGMEM to the line defining the file like this:

static const byte logo[] PROGMEM = {

The "logo" is the name of the image that was converted earlier. There is a routine in the "loop" that will create an array out of the BMP file that you pasted in with the modifications shown above. It will shift it up or down one row, each time through, according to the value of "c".

This next picture is of a Raspberry Pi and an Arduino Logo scrolling up and down on the LED sign.

```
// Version for sending data to 8 Parallel WS2812 strings
// by: BOB Davis
// Any BMP Viewer

// PORTD is Digital Pins 0-7 on the Uno change for other boards.
#define PIXEL_PORT  PORTD  // Port the pixels are connected to
#define PIXEL_DDR   DDRD   // Port the pixels are connected to
int textb=128;    // brightness of text
int maxled=180;   // maximum number of LED's in sign
byte sarray[180*3]; // three times max led
int imgwidth=90;  // image number of pixels wide
byte mbyte;
#include <avr/pgmspace.h>
// Set default color for letters
int red=10; int green=10; int blue=10;

// Send the next set of 8 WS2812B encoded bits to the 8 pins.
// The delay timing is for an Arduino UNO.
void sendBitX8( byte bits ) {
  PORTD= 0xFF;  // turn on
  PORTD= 0xFF;  // delay
  PORTD= 0xFF;  // delay
  PORTD= 0xFF;  // delay
  PORTD= 0xFF;  // delay (add more for faster processors)
  PORTD= bits;  // send data
```

```c
  PORTD= bits;  // delay
  PORTD= bits;  // delay
  PORTD= bits;  // delay
  PORTD= bits;  // delay
  PORTD= 0x00;  // Turn off;
}

int mask = 0x01;  // will shift mask to determine bit status
void sendPixelRow( byte mred, byte mgreen, byte mblue ) {
  // Send the bit 8 times down every row, each pixel is 8 bits/color
   for (int mbit=8; mbit>0; mbit--){
     if (green & mask<<mbit)sendBitX8( mgreen ); else sendBitX8( 0x00 );}
   for (int mbit=8; mbit>0; mbit--){
     if (red & mask<<mbit)sendBitX8( mred ); else sendBitX8( 0x00 ); }
   for (int mbit=8; mbit>0; mbit--){
     if (blue & mask<<mbit)sendBitX8( mblue ); else sendBitX8( 0x00 ); }
 }

// Generated by   : ImageConverter 565 Online
// Generated from : Logo-90.jpg
// bits per pixel: 24

static const byte Logo[] PROGMEM = {
// INSERT your converted BMP Here .
// It will be about twenty pages of numbers!
};

int mmax=maxled*3;  //maxled for all three colors
int tlevel = 120;  // trigger level
int mwide=imgwidth*3;
int c=0;     // controls scrolling
int dir =1; // controls direction of scrolling

void setup() {
  PIXEL_DDR = 0xff;    // Set all row pins to output
}

void loop() {
  // Clear array
  for (int k=0; k < mmax; k++)  {
```

```
  sarray[k] = 0x00;
 }
// Offset array
if (dir==1) {c=c+1;}
if (dir==0) {c=c-1;}
if (c==0) {dir=1;}
if (c==20) {dir=0;}
int coff = c*270; // offset * one row

 // Create top of array
 for (int k=0; k < imgwidth*3; k++)  {
  mbyte = pgm_read_byte_near(Logo+k+coff);
  if (mbyte>tlevel) {sarray[k]=sarray[k]+0x80;}
  mbyte = pgm_read_byte_near(Logo+k+imgwidth*3+coff);
  if (mbyte>tlevel) {sarray[k]=sarray[k]+0x40;}
  mbyte = pgm_read_byte_near(Logo+k+imgwidth*6+coff);
  if (mbyte>tlevel) {sarray[k]=sarray[k]+0x20;}
  mbyte = pgm_read_byte_near(Logo+k+imgwidth*9+coff);
  if (mbyte>tlevel) {sarray[k]=sarray[k]+0x10;}
  mbyte = pgm_read_byte_near(Logo+k+imgwidth*12+coff);
  if (mbyte>tlevel) {sarray[k]=sarray[k]+0x08;}
  mbyte = pgm_read_byte_near(Logo+k+imgwidth*15+coff);
  if (mbyte>tlevel) {sarray[k]=sarray[k]+0x04;}
  mbyte = pgm_read_byte_near(Logo+k+imgwidth*18+coff);
  if (mbyte>tlevel) {sarray[k]=sarray[k]+0x02;}
  mbyte = pgm_read_byte_near(Logo+k+imgwidth*21+coff);
  if (mbyte>tlevel) {sarray[k]=sarray[k]+0x01;}
  // create bottom half
  mbyte = pgm_read_byte_near(Logo+k+imgwidth*24+coff);
  if (mbyte>tlevel) {sarray[k+mwide]=sarray[k+mwide]+0x80;}
  mbyte = pgm_read_byte_near(Logo+k+imgwidth*27+coff);
  if (mbyte>tlevel) {sarray[k+mwide]=sarray[k+mwide]+0x40;}
  mbyte = pgm_read_byte_near(Logo+k+imgwidth*30+coff);
  if (mbyte>tlevel) {sarray[k+mwide]=sarray[k+mwide]+0x20;}
  mbyte = pgm_read_byte_near(Logo+k+imgwidth*33+coff);
  if (mbyte>tlevel) {sarray[k+mwide]=sarray[k+mwide]+0x10;}
  mbyte = pgm_read_byte_near(Logo+k+imgwidth*36+coff);
  if (mbyte>tlevel) {sarray[k+mwide]=sarray[k+mwide]+0x08;}
  mbyte = pgm_read_byte_near(Logo+k+imgwidth*39+coff);
  if (mbyte>tlevel) {sarray[k+mwide]=sarray[k+mwide]+0x04;}
  mbyte = pgm_read_byte_near(Logo+k+imgwidth*42+coff);
```

```
    if (mbyte>tlevel) {sarray[k+mwide]=sarray[k+mwide]+0x02;}
    mbyte = pgm_read_byte_near(Logo+k+imgwidth*45+coff);
    if (mbyte>tlevel) {sarray[k+mwide]=sarray[k+mwide]+0x01;}

  }
  red=textb; green=textb; blue=textb;
  cli();              // No time for interruptions!
  for (int l=0; l<mmax; l=l+3){    // send array
    sendPixelRow(sarray[l], sarray[l+1], sarray[l+2] );
  }
  sei();
  delay(200);         // Wait to display results
  return;
}
```

Bibliography

Running 8 LED strips in parallel using machine language to control the timing: (I use repeated commands as time delays instead)

https://wp.josh.com/2016/05/20/huge-scrolling-arduino-led-sign/

The WS2812 Specification sheet:

http://www.world-semi.com/

The MSGEQ7 Specification sheet:

http://www.mix-sig.com/index.php/msgeq7-seven-band-graphic-equalizer-display-filter

The HC06 Bluetooth Module Specifications:

https://www.olimex.com/Products/Components/RF/BLUETOOTH-SERIAL-HC-06/resources/hc06.pdf

Note: It is best to use a 5 volt adapter board with the Bluetooth module.

www.ingramcontent.com/pod-product-compliance
Lightning Source LLC
Chambersburg PA
CBHW062222220526
45471CB00009B/3308